# Environmental Toxin Series 3

Editors-in-Chief: S. Safe and O. Hutzinger

# Environmental Toxin Series

Editors-in-Chief: S. Safe and O. Hutzinger

*Volume 1* Polychlorinated Biphenyls (PCBs):
Mammalian and Environmental Toxicology
S. Safe (Ed.)

*Volume 2* Cadmium

M. Stoeppler and M. Piscator (Eds.)

*Contents*:
I. Toxicity, Carcinogenicity, Animal Experiments
II. Epidemiology
III. Cadmium in the Environment
IV. Methodology and Quality Assessment

S. Safe, O. Hutzinger, T. A. Hill (Eds.)

# Polychlorinated Dibenzo-p-dioxins and -furans (PCDDs/PCDFs): Sources and Environmental Impact, Epidemiology, Mechanisms of Action, Health Risks

With 8 Figures and 32 Tables



Springer-Verlag Berlin Heidelberg NewYork
London Paris Tokyo Hong Kong Barcelona

*Volume Editors*

Prof. Dr. S. Safe
Texas A&M University, College of Veterinary Physiology and Pharmacology, College Station, TX 77843-4466, USA

Prof. Dr. O. Hutzinger
Institute of Ecological Chemistry and Geochemistry, University of Bayreuth, P.O. Box 101251, 8580 Bayreuth, FRG

T.A. Hill
Consultants in Epidemiology and Occupational Health, 2428 Wisconsin Avenue, NW Washington, DC 20007, USA

ISBN 3-540-15552-X Springer-Verlag Berlin Heidelberg New York
ISBN 0-387-15552-X Springer-Verlag New York Berlin Heidelberg

Library of Congress Cataloging-in-Publication Data
Polychlorinated dibenzo-*p*-dioxins and -furans (PCDDs/PCDFs): sources and environmental impact, epidemiology, mechanisms of action, health risks/S. Safe, O. Hutzinger, T.A. Hill (eds.).
(Environmental toxin series; 3)
ISBN 0-387-15552-X (U.S.: alk. paper)
1. Polychlorinated dibenzodioxins – Toxicology. 2. Polychlorinated dibenzodioxins – Environmental aspects. 3. Polychlorinated dibenzofurans – Toxicology. 4. Polychlorinated dibenzofurans – Environmental aspects. I. Safe, S. II. Hutzinger, O. III. Hill, Thomas A. IV. Series: Environmental toxin series; v. 3. RA1242.P72P65 1990
615.9′511 – dc20 90-9703


Printed in Germany



Typesetting: Macmillan India Ltd., Bangalore-25
Offsetprinting: Mercedes-Druck, Berlin; Bookbinding: B. Helm, Berlin 2152/3020-543210 – Printed on acid-free paper

# Editorial

The concern about environmental toxins is constantly increasing, as is the need for sound scientific information. The Environmental Toxin Series is dedicated to the publication of comprehensive reviews and monographs on compounds or classes of chemicals which are of importance in environmental toxicology. The series is designed to serve as a background of information for scientific investigation as well as risk analysis and political decision making. The main aim of the series is to describe in as complete a way as possible all potentially hazardous chemicals from the point of view of chemistry, ecology, toxicology, risk analysis and regulatory implications. From time to time conference proceedings on important and urgent topics will be included in the series. We thank the members of the editorial board for their enthusiastic support.

*S. Safe and O. Hutzinger*

ginia College of Medicine, P.O.Box 267, Richmond, VA 23298-0001, USA

Dr. A.W. Hayes
RJR Nabisco, Inc., Bowman Gray Technical Center, Winston-Salem, NC 27109, USA

Prof. Dr. M.A. Hayes
Department of Pathology, University of Guelph, Guelph, Ontario N1G 2W1, CND

Dr. G.W. Ivie
Agricultural Research Service, U.S. Department of Agriculture, College Station, TX 77841, USA

Dr. R. Koch
Academy of Sciences, Institute of Geography and Geo-Ecology, Rudower Chaussee 5, 1199 Berlin, GDR

Dr. H.J. Lewerenz
Academy of Sciences of the GDR, Central Institute of Nutrition, Artur-Scheunert-Allee, 1505 Bergholz-Rehbrücke, GDR

Dr. E. Löser
Bayer AG, Institute of Industrial Toxicology, P.O. Box 101709, 6500 Wuppertal 1, FRG

Dr. J.D. McKinney
Division of Molecular Biophysics N.I.E.H.S., P.O. Box 12233, Research Triangle Park, NC 27709, USA

Prof. Dr. A. Parkinson
Department of Pharmacology, Toxicology and Therapeutics, University of Kansas, Medical Center, 39th St., Kansas City, KS 66103, USA

Dr. T.D. Phillips
Texas A&M University, Department of Public Health, College of Veterinary Medicine, College Station, TX 77843, USA

Dr. I.G. Sipes
Department of Pharmacology and Toxicology, University of Arizona, College of Pharmacy, Tucson, AZ 85721, USA

Dr. J. Thies
Bayer AG, Analytical Laboratory Center, OAL, 5090 Leverkusen, FRG

Dr. H.R. Witschi
Oak Ridge National Laboratory, Biology Division, P.O. Box Y, Oak Ridge, TN 37831, USA

# Contents

1 PCDDs and PCDFs: Sources and Environmental Impact
S. Safe and O. Hutzinger . . . . . . . . . . . . . . . . . . . . 1

2 UAREP-Report on Health Aspects of Polychlorinated Dibenzo-*p*-dioxins (PCDDs) and Polychlorinated Dibenzofurans (PCDFs). . . . . . . . . . . . . . . . . . . . . . . . 23

2.1 Epidemiology
A.H. Smith, L.T. Kurland and S. Shindell . . . . . . 27

2.2 Mechanism of Action
S.H. Safe, T. Gasiewicz and J.P. Whitlock, Jr. . . . . 61

2.3 Characterization of Human Health Risks
C.O. Schulz, D.R. Brown and I.C. Munro . . . . . . 93

3 Subject Index . . . . . . . . . . . . . . . . . . . . . . . . . . . . 141

# 1 PCDDs and PCDFs: Sources and Environmental Impact

**S. Safe[1] and O. Hutzinger[2]**

1 Introduction . . . . . . . . . . 1
2 Sources of PCDDs and PCDFs . . . . . . . . . . 2
2.1 Chemical Sources . . . . . . . . . . 4
2.2 Pulp and Paper Mills and Paper/Board Products . . . . . . . . . . 7
2.3 Combustion Sources . . . . . . . . . . 8
3 PCDDs and PCDFs – Environmental Impact . . . . . . . . . . 11
3.1 Atmospheric Levels . . . . . . . . . . 11
3.2 Sediments, Sludge . . . . . . . . . . 12
3.3 Environmental Levels . . . . . . . . . . 12
3.4 Yusho and Yu-Cheng Incidents . . . . . . . . . . 16
4 References . . . . . . . . . . 17

## 1 Introduction

Polychlorinated dibenzo-*p*-dioxins (PCDDs or polyCDDs) and polychlorinated dibenzofurans (PCDFs or polyCDFs) are members of the family of halogenated aromatic chemicals which includes the polychlorinated biphenyls (PCBs), naphthalenes, terphenyls, azo and azoxybenzenes, quaterphenyls and their brominated and mixed chloro/bromo analogs. The PCDDs and PCDFs can be substituted with 1–8 chlorine atoms and there are 75 and 135 possible isomers and congeners. With the exception of the fully chlorinated octaCDD and octaCDF, there are multiple isomers within each group as illustrated in Fig. 1.

The PCDFs, PCDDs and structurally-related halogenated aryl hydrocarbons exhibit a number of common properties. For example, the PCDDs and PCDFs are highly lipophilic molecules and are generally resistant to chemical breakdown by acids, bases or by oxidation, reduction and hydrolysis. These physicochemical properties were important properties which contributed to the diverse number of industrial applications of PCBs. In addition, the degree of

[1] Veterinary Physiology and Pharmacology, Texas A&M University, College Station, Texas 77843, USA
[2] Department of Ecological Chemistry & Geochemistry, University of Bayreuth, Postfach 3008, D-8580, Bayreuth, FRG

Environmental Toxin Series, Vol. 3

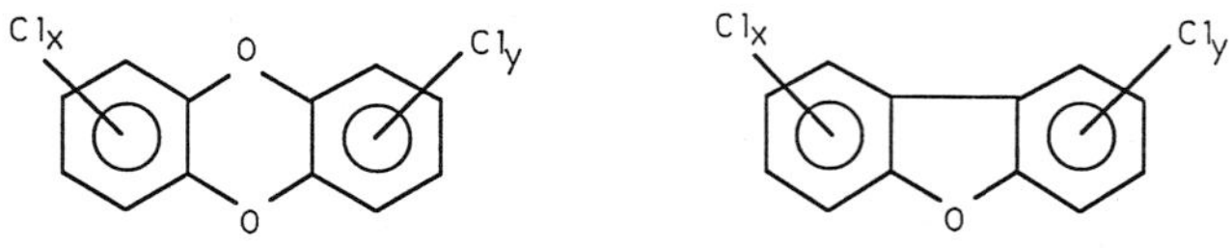


| No. of Cl groups | No. of Isomers | No. of Isomers |
|---|---|---|
| 1 | 2 | 4 |
| 2 | 10 | 16 |
| 3 | 14 | 28 |
| 4 | 22 | 38 |
| 5 | 14 | 28 |
| 6 | 10 | 16 |
| 7 | 2 | 4 |
| 8 | 1 | 1 |
| TOTAL | 75 | 135 |

**Fig. 1.** PCDDs and PCDFs – multiplicity of isomers and congeners

lipophilicity and stability of PCDDs and PCDFs also increases with increasing ring chlorination. Not surprisingly, these physicochemical properties contribute to the environmental stability of these compounds which have been detected in almost every component of the global ecosystem. Moreover, the lipophilic nature of these hydrocarbons has resulted in their bioaccumulation in the food chain and PCDDs and PCDFs have been reported in fish, wildlife and human adipose tissue, milk and serum.

The widespread identification of trace levels of PCDDs and PCDFs in human populations has engendered considerable public and regulatory concern over the potential adverse human health impacts of these pollutants. In addition, this concern has been heightened by several accidental and occupational exposures to PCDDs and PCDFs. These incidents have been thoroughly documented and are summarized in this book. Figure 1 shows that there are a large number of PCDD and PCDF congeners, and humans are exposed to complex mixtures of these compounds. However, in exposures associated with 2,4,5-trichlorophenol and its derived products [e.g. 2,4,5-trichlorophenoxyacetic acid (2,4,5-T) or Agent Orange], the major contaminant has been identified as 2,3,7,8-tetrachlorodibenzo-*p*-dioxin (TCDD). 2,3,7,8-TCDD is the most toxic halogenated aromatic compound and has been utilized as a prototype for investigating the toxicity, carcinogenicity and mechanisms of action of halogenated aryl hydrocarbons.

2,3,7,8-TCDD exhibits exquisitely high toxicity in some laboratory animal species (e.g. $LD_{50}$ guinea pigs – < 2 μg/kg) and is a potent rodent carcinogen. For example, hyperplastic liver nodules were observed in female Sprague-Dawley rats fed 2,3,7,8-TCDD at a dose of 0.01 μg/kg/day. Thus early studies on the sources and environmental impact of PCDDs and PCDFs primarily focused on the specific detection and quantitation of 2,3,7,8-TCDD. However, it is now recognized that the 2,3,7,8-substituted tetra- to hexachlorinated dibenzo-*p*-dioxins and dibenzofurans (Figs. 2 and 3) are also highly toxic and many current analytical studies now report residue levels of these compounds and the less toxic hepta- and octachlorinated isomers.

## 2 Sources of PCDDs and PCDFs

In the search for chlorinated, brominated and mixed chloro/bromo dibenzo-*p*-dioxin and dibenzofurans several processes have been listed according to their potential as sources of these compounds [1]. These include industrial sources associated with the production or use of chlorinated phenols and their derived products, chlorinated benzenes, brominated flame retardants, aliphatic organochlorines, other organochlorinated compounds, inorganics which contain chlorine and chlorine-containing catalyst or solvents [2]. In addition, several other potential sources of PCDDs and PCDFs include the pulp and paper industry, metallurgical processes, and procedures used for the reactivation of granular carbon.

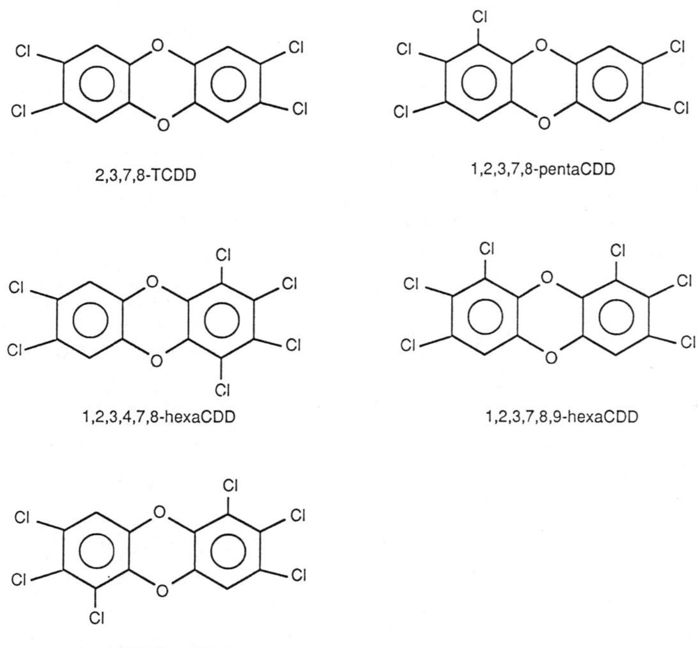


**Fig. 2.** Structures of the most toxic tetra- hexaCDDs

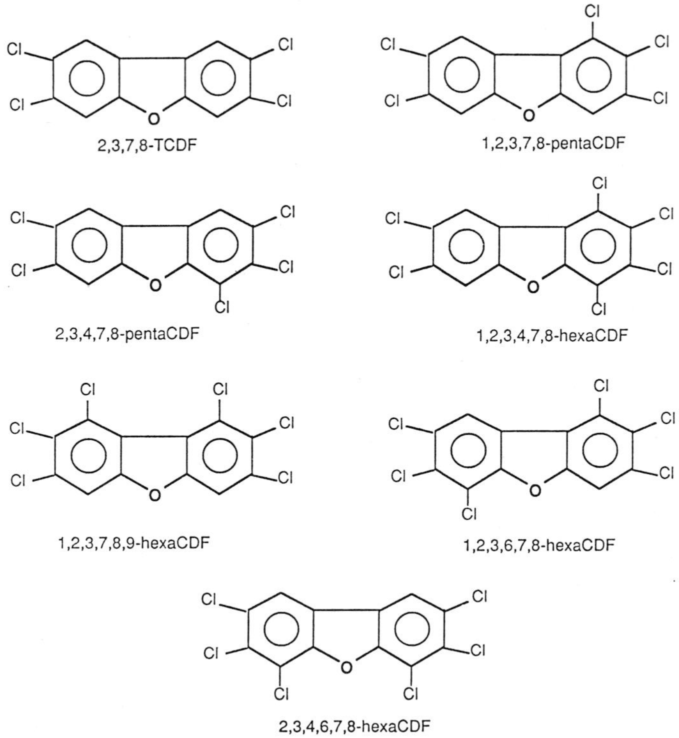


**Fig. 3.** Structures of the most toxic tetra- hexaCDFs

## *2.1 Chemical Sources*

Commercial chlorinated phenols are industrial compounds which have been widely used as bactericides, slimicides, wood preservatives and fungicides and as precursors in the production of a variety of other industrial chemicals including phenoxy herbicides, hexachlorophene and chlorinated diphenylether herbicides. PCDFs and PCDDs are routinely identified as impurities in chlorinated phenols and their derived products. Table 1 summarizes the concentrations of these impurities in several commercial preparations. The levels of PCDDs and PCDFs in the commercial products are highly variable and depend on several factors including the reaction conditions (particularly temperature) and the "cleanup" procedures. Since PCDDs and PCDFs are neutral compounds and chlorinated phenols are weak acids, it is possible to solvent extract these impurities from the sodium chlorophenates. The structures of individual PCDD and PCDF isomers and congeners in chlorinated phenols are related to the degree of chlorination of the phenol. For example, the PCDD and PCDF impurities found in pentachlorophenol preparations (Table 1) are primarily the higher chlorinated octa and heptaCDDs and CDFs whereas the lower chlorin-

**Table 1.** PCDD and PCDF levels in chlorinated phenols and their derived products (ppb)

| Congener | Dowicide G[a] | 2,4,5-T Preparations |
|---|---|---|
| 2,3,7,8-TCDD | 0.23 | $0.4–6.1 \times 10^3$ [b] |
| 1,2,3,7,8-pentaCDD | 18.2 | |
| 1,2,3,4,7,8-hexaCDD | 28.3 | |
| 1,2,3,6,7,8-hexaCDD | $2.03 \times 10^3$ | |
| 1,2,3,7,8,9-hexaCDD | 282 | |
| 1,2,3,4,6,7,8-heptaCDD | $14.8 \times 10^3$ | |
| octaCDD | $41.6 \times 10^3$ | |
| 2,3,7,8-TCDF | 1.8 | 60[b] |
| 1,2,3,7,8-pentaCDF | 8.2 | |
| 2,3,4,7,8-pentaCDF | 6.6 | |
| 1,2,3,4,7,8-hexaCDF | 48 | |
| 1,2,3,6,7,8-hexaCDF | 69 | |
| 1,2,3,7,8,9-hexaCDF | $< 1$ | |
| 2,3,4,6,7,8-hexaCDF | 87 | |
| 1,2,3,4,6,7,8-heptaCDF | 699 | |
| 1,2,3,4,7,8,9-heptaCDF | 675 | |
| octaCDF | $37.2 \times 10^3$ | |

[a] Pentachlorophenol Ref. [3]; [b] Ref. [116]

ated phenols contain the lower chlorinated CDDs and CDFs [3, 4]. In summary, PCDDs and PCDFs can be formed as byproducts in the synthesis of the following technical chlorophenols: 2-, 3- and 4-chlorophenol, 2,4-dichlorophenol, 2,4,5-, 2,4,6- and 3,4,5-trichlorophenol, 2,3,4,5- and 2,3,4,6-tetrachlorophenol and pentachlorophenol. The factors which enhance PCDD/PCDF formation include alkaline conditions, high temperatures, ultraviolet irradiation and other radical initiators. The formation of PCDDs and PCDFs from other chlorinated aromatics is enhanced when production or purification is carried out under alkaline conditions, when the precursors contain an aromatic C-O moiety and when temperatures exceed 150 °C.

The detection of PCDFs in commercial PCB mixtures constituted one of the first in a series of reports which identified PCDFs as contaminants in commercial products [5–12]. The high resolution GC-MS analysis of the PCDFs in several commercial PCBs has been reported by Wakimoto et al. [13], and Table 2 summarizes the levels of the individual 2,3,7,8-substituted congeners and the total tetra- octaCDFs. It is evident that the commercial products are highly complex mixtures of isomers and congeners and that the overall PCDF levels vary from 0.598–26.0 ppm in Kanechlor 400 and Aroclor 1242, respectively. There were considerable variations in the concentration of the total PCDFs and the individual 2,3,7,8-substituted PCDFs between the commercial PCBs; however, it was evident that there was an increase in the levels of the higher chlorinated PCDFs as the % chlorine content of the PCB preparations increased. The concentrations of PCDDs in the commercial PCBs are not significant. Another recent study has reported the levels of the 3 highly toxic PCDFs which occur in

**Table 2.** PCDFs in commercial PCBs, (ppb) [13]

| Congener | KC-300 | KC-400 | KC-500 | KC-600 | Ar-1242 | Ar-1248 | Ar-1254 | Ar-1260 |
|---|---|---|---|---|---|---|---|---|
| 2,3,7,8-TCDF[a] | 502 | 1680 | 284 | 116 | 93 | 314 | 165 | 129 |
| 1,2,3,7,8-pentaCDF[b] | 68 | 230 | 69 | 24 | 3 | 53 | 128 | 140 |
| 2,3,4,7,8-pentaCDF | 295 | 679 | 429 | 165 | 19 | 118 | 262 | 166 |
| 1,2,3,4,7,8-hexaCDF[c] | 275 | 374 | 553 | 223 | < 1 | 18 | 392 | 566 |
| 1,2,3,6,7,8-hexaCDF | 121 | 163 | 218 | 91 | < 1 | 11 | 163 | 210 |
| 1,2,3,7,8,9-hexaCDF | 7 | 6 | 11 | 4 | < 1 | < 1 | 11 | 34 |
| 2,3,4,6,7,8-hexaCDF | 35 | 33 | 50 | 41 | < 1 | < 1 | 24 | 27 |
| 1,2,3,4,6,7,8-heptaCDF | 104 | 128 | 165 | 476 | < 1 | < 5 | 92 | 109 |
| 1,2,3,4,7,8,9-heptaCDF | 59 | 38 | 143 | 104 | < 1 | < 5 | 69 | 180 |
| tetraCDF | 4390 | 18,800 | 1490 | 557 | 489 | 250 | 659 | 502 |
| pentaCDF | 1610 | 5420 | 2790 | 857 | 87 | 1090 | 2000 | 1120 |
| hexaCDF | 1070 | 1320 | 2120 | 1030 | 21 | 73 | 1280 | 1880 |
| heptaCDF | 301 | 273 | 637 | 1080 | — | — | 273 | 1980 |
| octaCDF | 172 | 200 | 163 | 1880 | < 10 | < 16 | < 22 | 2020 |
| Total PCDFs | 7540 | 26,000 | 7200 | 5410 | 598 | 3730 | 4220 | 7500 |

[a] Includes the 2,3,4,8-TCDF congener; [b] includes the 1,2,3,4,8-pentaCDF congener; [c] includes the 1,2,3,4,7,9-hexaCDF congener

commercial PCBs, namely 2,3,7,8-TCDF, 2,3,4,7,8-pentaCDF and 1,2,3,7,8,9-hexaCDF [14]. Their high resolution analytical results indicated that the mean levels of these compounds in Aroclors 1010, 1242, 1254 and 1260 were: 2,3,7,8-TCDF 0.24, 37.0, 23.8 and 20.9 ng/g, respectively; 2,3,4,7,8-pentaCDF, 0.74, 60.0, 146 and 79.6 ng/g, respectively; 1,2,3,7,8,9-hexaCDF, 0.02, 0.77, 29.1 and 49.8 ng/g, respectively. A comparison of the results reported by Wakimoto et al. [13] and Brown et al. [14] indicated that there were considerable variations in the concentrations of specific congeners. For example, in one study the concentrations of 2,3,4,7,8-pentaCDF in Aroclors 1242, 1254 and 1260 were 60, 146 and 79.6 ng/g [14]; whereas, the levels were 19, 262 and 166 ng/g, respectively, in the second report [13]. These differences were not surprising since the PCDF levels in a sample of the same commercial PCB can vary considerably, and this is related in part to the dibenzofuran concentrations in the commercial biphenyl feedstock used in the chlorination process.

The Yusho and Yu-Cheng "PCB poisonings", which were caused by PCB-contaminated rice oil, are now presumed to be due to the PCDFs present in these oils, and the results in Table 3 summarize the relative PCDF concentrations in the contaminated oils and the PCDF/PCB ratios. These ratios were 0.22 and 0.13 for the Yusho and Yu-Cheng samples; whereas, the ratios of PCDFs/PCBs in the commercial Kanechlors 400 and 500 were 0.0018 and 0.0004, respectively [15].

PCDDs and PCDFs have been detected as by-products in other commercial chemicals, and these include: hexachlorobenzene [16], chlorides of iron, aluminum and copper [1], and technical hexachlorocyclohexanes [17].

## 2.2 *Pulp and Paper Mills and Paper/Board Products*

The release of PCDFs and PCDDs from the pulp and paper industry have recently been characterized as a new environmental source for these contaminants. The chlorination of naturally-occurring phenolic compounds, such as

**Table 3.** Comparation PCDF levels in Yusho and Yu-Cheng oils [15]

| Analyte | Concentration (ppm) Yusho oil | Yu-Cheng oil |
|---|---|---|
| Unknown tetraCDF | 0.65 | 0.05 |
| 2,3,7,8-tetraCDF | 0.20 | trace |
| 2,3,4,7,8-pentaCDF | 0.70 | 0.02 |
| 2,3,4,6,7-pentaCDF | 0.35 | 0.02 |
| 1,2,3,4,7,8-1,2,3,6,7,8-hexaCDF | 0.12 | 0.01 |
| Total | 2.02 | 0.10 |
| PCBs | 920 | 8 |
| PCDFs/PCBs | 0.22 | 0.13 |

lignin and other plant constituents, can result in the formation of chlorophenols. In addition, chlorinated phenoxyphenols are formed during the chlorobleaching of wood pulp in the paper industry and both the chlorinated phenols and phenoxyphenols are precursors of PCDDs and PCDFs [18–20]. Analysis of grab samples and sediments in the vicinity of a pulp and paper mill producing bleached softwood kraft pulp identified ppt levels of the 2,3,7,8-substituted PCDDs and PCDFs [21]. Subsequent analyses of sludge from pulp and paper mills and samples of pulp and process liquors from pulp and paper mills have confirmed the presence of 2,3,7,8-substituted PCDFs and PCDDs at the ppt levels [22–25]. Beck and coworkers [26] have also analysed several finished paper products including newsprint, laboratory filter paper, coffee filters, cosmetic tissue and recycled paper for PCDDs and PCDFs. Although tetra-octachlorinated dibenzo-*p*-dioxins and dibenzofurans were identified in most of the paper products, the dominant compounds were the more toxic 2,3,7,8-substituted congeners, and the highest overall concentrations were detected in the recycled paper. A similar spectrum of PCDDs and PCDFs was detected in samples of pulp and board products [27] and it is generally acknowledged that the PCDDs and PCDFs are formed during the bleaching process.

### *2.3 Combustion Sources*

Olie, Vermeulen and Hutzinger [28] first reported the detection of complex mixtures of PCDDs and PCDFs as trace components of fly ash and flue gas emissions from municipal waste incinerators in the Netherlands. Subsequent research in several laboratories has identified PCDDs and PCDFs as combustion products from municipal waste incinerators throughout the world [21, 29–31], and it was apparent that the absolute levels of the PCDDs and PCDFs

**Table 4.** Typical ranges of PCDD/PCDF-contents in flue gases and residues from municipal waste combustors

| Sources | Compounds | Concentrations |
|---|---|---|
| Flue Gas: | 2,3,7,8-TCDD | 0.05–0.07 [ng/m$^3$] |
| | Σ tetraCDDs | 4–25 [ng/m$^3$] |
| | Σ pentaCDDs | 20–145 [ng/m$^3$] |
| | octaCDD | 4–49 [ng/m$^3$] |
| | 2,3,7,8-TCDF | 0.9–6 [ng/m$^3$] |
| | Σ tetraCDFs | 22–144 [ng/m$^3$] |
| | Σ pentaCDFs | 91–361 [ng/m$^3$] |
| | $Cl_8DF$ octaCDF | 1–8 [ng/m$^3$] |
| Clean Gas Dust: | PCDDs/PCDFs | 0.25–3 [ng/g] |
| Filter Dust: | PCDDs/PCDFs | 1.2–1.4 [ng/g] |
| Slag: | PCDDs/PCDFs | < 20 [pg/g] |

formed during the incineration process were dependent on the incinerator design and operating conditions [32, 33]. Table 4 summarizes the range of PCDD and PCDF concentrations in flue gases and residues from municipal waste combustors [34]. The stack gas concentrations are the sum of the PCDDs and PCDFs found in the particulate and gaseous phases, and the particulate content of the stack gases ranged from 10 to 20 mg/m$^3$. Several studies have also confirmed the formation of PCDDs and PCDFs in home heating systems which use both coal and wood [35–38]. Table 5 summarizes the relative concentrations of PCDDs and PCDFs in soot from municipal waste incinerators and from home heating incinerators, and it is clear that the higher levels were observed in the municipal waste incinerators. The "Trace Chemistries of Fire Hypotheses" proposed by scientists at the Dow Chemical Co. [29] suggested the PCDFs and PCDDs can arise from diverse combustion processes which include "natural" forest fires and volcanic eruptions. It is apparent from dated aquatic sediment profiles that there were time-dependent increases in the concentrations of PCDDs and PCDFs and this correlated with the increased production and use of chlorinated industrial compounds [39–43]. These results suggest that the industrial production of halogenated organic chemicals and their incineration contribute more to the accumulation of PCDDs and PCDFs in sediments than do natural combustion processes. PCDFs and PCDDs have also been detected as by products of the combustion of numerous fuel sources including coal, wood and oil [37, 44, 45], and sewage sludge [46].

Several laboratories have investigated the formation of PCDDs and PCDFs from the combustion or pyrolysis of individual classes of chlorinated organic compounds which may be precursors. Laboratory combustion of PCBs can give relatively high yields of PCDFs [47–49]. Not surprisingly, the major toxicological concerns associated with PCB fires have been the formation of PCDFs

**Table 5.** Concentration of PCDD/PCDF in the soot of home heating (ppb)

| Compounds | MWI Fly Ash | Home Heating Soot Oil | Soot Coal/Wood |
|---|---|---|---|
| tetraCDDs | 19.06 | 3.90 | 1.54 |
| pentaCDDs | 37.25 | 0.41 | 5.13 |
| hexaCDDs | 115.49 | 3.06 | 5.20 |
| heptaCDDs | 275.88 | 1.37 | 2.76 |
| octaCDD | 598.69 | 1.05 | 2.37 |
| tetraCDF | 79.45 | 28.91 | 50.80 |
| pentaCDF | 120.30 | 16.55 | 30.04 |
| hexaCDF | 116.34 | 6.24 | 11.67 |
| heptaCDF | 108.24 | 1.78 | 3.23 |
| octaCDF | 42.90 | 0.33 | 0.53 |
| 2,3,7,8-tetraCDD | 0.60 | 0.10 | 0.21 |
| 2,3,7,8-tetraCDF | 2.47 | 1.07 | 1.92 |

[31, 50–53]. It has been reported that PCDFs and PCDDs are formed from the combustion of chlorophenates [54, 55] and chlorinated phenol-derived products [56], polyvinyl chloride and chlorinated alkanes [57, 58] and chlorinated benzenes [59]. Halogenated aromatic compounds (including several brominated chemicals) have been identified in the emissions from automobiles using leaded gasolines [60]. Dibromoethane and dichloroethane are added to leaded gasolines and act as scavengers to prevent the deposition of lead compounds in engines. These compounds are not added to unleaded gasoline and, not surprisingly, no halogenated aromatics were emitted from automobiles using unleaded gasoline (Table 6) [61].

Several studies have speculated on the mechanisms involved in the combustion-derived formation of PCDDs and PCDFs [62–66]. It is apparent that under the appropriate experimental conditions PCDDs and PCDFs can be formed from simple chemical precursors. For example, in a flow reactor experiment using methane gas, hydrochloric acid and oxygen, over 300 compounds were formed, and these include alkanes, alkenes, alkynes, aromatic hydrocarbons and chlorinated aromatic hydrocarbons including PCDDs and PCDFs [67]. Since PCBs and chlorinated phenols are utilized as synthetic precursors in the synthesis of PCDDs and PCDFs [47–49, 66], these compounds no doubt play an important role in the "in situ" formation of PCDDs and PCDFs from diverse combustion processes. In a recent study it was reported that over a wide range of temperatures PCDDs were produced from the reaction of [$^{13}C_6$]-pentachlorophenol on various fly ash particulates [66]. In addition, Rappe and coworkers have compared the GC-MS profiles of the PCDDs and PCDFs derived from several combustion sources [21, 61, 68]. These studies showed that for samples from several different combustion sources there was a remarkable qualitative similarity between their GC-MS fragmentograms for various isomer groups. These results suggest that there are common pathways for the formation of PCDDs and PCDFs during the combustion or pyrolysis of organic material.

**Table 6.** PCDDs and PCDFs in car exhausts (ng/24.8 km) [61]

| Compounds | Cars running on nonleaded gasoline (n = 2) | Cars running on leaded gasoline (n = 4) |
|---|---|---|
| 2,3,7,8-TCDF | ND | 0.6–13 |
| Σ tetraCDFs | ND | 10–200 |
| 2,3,7,8-TCDD | ND | < 0.05–0.3 |
| Σ tetraCDDs | ND | 3–90 |
| 1,2,3,7,8-pentaCDF | ND | 0.5–4.6 |
| 2,3,4,7,8-pentaCDF | ND | 0.3–3.9 |
| Σ pentaCDFs | ND | 5–46 |
| 1,2,3,7,8-pentaCDD | ND | 0.5–3.5 |
| Σ pentaCDDs | ND | 6–98 |
| TCDD equivalents | ND | 0.8–13 |

## 3 PCDDs and PCDFs – Environmental Impact

The PCDDs and PCDFs which are present in industrial chemicals and, in waste from pulp and paper facilities and which are formed during combustion processes constitute the major inputs of these compounds into the global environment. The relative contributions of these and other, as yet unidentified, sources of PCDDs and PCDFs to the total environmental inputs of these compounds are currently unknown. Complex mixtures of PCDDs and PCDFs have been identified in diverse environmental extracts at sub ppb levels and this is consistent with the relatively low concentrations of these compounds in their respective sources.

### *3.1 Atmospheric Levels*

Rappe et al. [69] have recently reported the identification and quantitation of PCDFs and PCDDs in atmospheric samples. Table 7 summarizes the atmospheric levels in several locations around the city of Hamburg, and it was apparent that the highest levels were observed in the vicinity of facilities which generate combustion-derived emissions of PCDFs and PCDDs. Nakano et al. [70] also reported that the average PCDD and PCDF levels around Kobe, Japan, were 8.6 and 8.8 pg/m$^3$, respectively; and these values were comparable to those reported for Hamburg air samples. It was apparent from the analysis of the air samples that the pattern of PCDFs and PCDDs was similar to patterns

**Table 7.** PCDFs and PCDDs from air samples in Hamburg (pg/m$^3$) [69]

| Sample* Designation | TCDF/TCDD tetraCDF TCDF | pentaCDF pentaCDD | hexaCDF hexaCDD | heptaCDF heptaCDD | OCDF OCDD |
|---|---|---|---|---|---|
| 1 | ND/ND | ND/ND | ND/ND | ND/0.28 | ND/0.37 |
| 2 | 0.36/0.01 | 0.51/0.05 | 0.18/0.74 | 0.1/0.60 | ND/0.37 |
| 3 | ND/ND | ND/ND | ND/ND | ND/ND | ND/0.27 |
| 4 | 1.6/0.37 | 1.0/0.37 | 3.2/1.0 | 0.47/1.5 | 0.19/4.2 |
| 5a | 0.58/0.35 | 0.67/ND | 0.68/0.93 | 1.0/7.9 | 0.14/1.3 |
| 5b | 3.3/1.5 | 10.0/5.0 | 9.5/24.4 | 5.1/15.4 | 7.0/40 |
| 5c | 1.59/0.88 | 1.34/0.41 | 1.27/2.60 | 2.40/5.80 | < 2.0/7.90 |
| 6a | 4.9/0.21 | 5.0/2.4 | 2.2/5.3 | 2.0/5.3 | 0.78/7.4 |
| 6b | 4.03/1.45 | 3.90/0.49 | 2.72/3.40 | 3.60/4.80 | < 2.40/5.40 |
| 7a | 2.8/0.05 | 2.0/0.35 | 0.75/0.24 | 0.20/1.5 | < 0.97/3.7 |
| 7b | 18.7/6.53 | 11.6/1.19 | 5.1/7.30 | 4.20/7.70 | < 0.96/7.70 |
| 8a | 3.5/0.23 | 3.6/2.5 | 2.0/7.8 | 1.9/3.4 | < 1.3/6.3 |
| 8b | 6.2/0.22 | 4.1/1.3 | 1.1/2.7 | 1.9/3.4 | < 1.0/6.4 |

* 1, rural/suburban area; 2, suburban area; 3, landfill site near flare gas site; 4, residential area near 3; 5, residential near 3 and west of a copper smelter; 6, industrial area (near motorway and smelter); 7, industrial area downwind from MSW incinerator; 8, motorway tunnel

observed from combustion emission sources. A recent report by Swerev and Ballschmiter [71] indicated that for air particulates from Ulm and two National Bureau of Standards particulate samples the pattern of PCDDs and PCDFs suggests that there was input from combustion-derived processes and commercial products (PCBs and pentachlorophenol). Hutzinger and coworkers have utilized waxy plant surfaces as a biomonitor for atmospheric PCDFs and PCDDs and their results show that PCDFs/PCDDs levels were approximately ten times higher in spruce needles collected from industrial areas compared to levels in needles from trees located in "clean air" areas [72]. Moreover, the pattern of PCDDs/PCDFs observed in the pine needle extracts was similar to the distribution of these compounds found in combustion-derived extracts. It was assumed that the background levels of PCDDs and PCDFs in pine needles from clean areas were due to other combustion sources which include home heating units. Clearly, this new method of "in situ" biomonitoring provides a novel approach for monitoring the atmospheric levels of lipophilic chemical pollutants.

### *3.2 Sediments, Sludge*

Hites and coworkers have analyzed sediment cores from several lakes for PCDDs and PCDFs. The depth vs chemical concentration profiles suggest that the levels of these compounds have dramatically increased since the 1940s and this corresponds to the increased production, use and disposal of chlorinated compounds [39–43]. The results also suggest that combustion-derived PCDFs and PCDDs were the major contributors to the increased sediment levels of these environmental contaminants [73]. Evers et al. [74] have reported spatial variations in the distribution of PCDDs and PCDFs in Rhine River sediments, and this can be attributed not only to atmospheric inputs but also to point source pollution problems. PCDDs and PCDFs have also been identified as contaminants of sewage sludge [75] and in the top soil from a contaminated refuse dump [76].

### *3.3 Environmental Levels*

Studies on the environmental levels of PCDFs and PCDDs initially focused on the detection and quantitation of 2,3,7,8-TCDD which was identified at the ppt level in freshwater fish [77, 78]. However, in recent years numerous reports on the high resolution analysis of PCDDs and PCDFs have confirmed their widespread distribution in environmental samples. Rappe and coworkers [79] identified PCDFs in fat samples from a snapping turtle (Hudson River) and grey seals from the Gulf of Bothnia, and subsequent research has identified PCDDs and PCDFs in fish and wildlife samples throughout the global aquatic and marine environments. For example, PCDDs and PCDFs have been detected in Great Lakes fish and birds [80–82], fish and fish-eating birds from the Netherlands [83], arctic seals from Spitzbergen [84], blue mussel, river and estuary sediments from Osaka Bay [85, 86], and in fish from inland lakes in Sweden and from the Baltic Sea [87]. It was apparent that the 2,3,7,8-substituted PCDDs and PCDFs were the major congeners identified in these samples; however, the relative concentrations of the individual components were highly variable. This variability is no doubt related to the composition of the sources of the PCDD and PCDF pollutants.

In a recent study, the high resolution analyses of PCDDs and PCDFs in fish samples from several of the Great Lakes has been reported [88]. It was apparent from the results that the PCDD and PCDF levels were lowest in fish from Lake Superior (Table 8) and Lake Erie and highest in fish from Lake Ontario (Table 9). The results also showed that there were higher levels of total PCDFs than total PCDDs in all the fish extracts from Lake Erie, Lake Superior, Lake Huron, Lake Michigan and Lake St. Clair. In addition, the predominant PCDF congeners were 2,3,7,8-TCDF, 1,2,3,7,8-pentaCDF and 2,3,4,7,8-pentaCDF, and the predominant PCDDs were 2,3,7,8-TCDD, 1,2,3,7,8-pentaCDD and

**Table 8.** PCDDs and PCDFs in Lake Superior fish – GC-MS and bioassay results (parts per trillion)

| | Sample Code | | | | |
|---|---|---|---|---|---|
| | 84-02-101 | 84-02-104 | 84-02-105 | 84-02-103 | 84-02-102 |
| A. Congener | | | | | |
| 2,3,7,8-tetraCDF | 19 | 5.7 | 22 | 17 | 15 |
| 1,2,3,7,8-pentaCDF | 1.3 | 0.6 | 2.3 | 1.8 | 2.4 |
| 2,3,4,7,8-pentaCDF | 2.7 | 1.0 | 3.0 | 3.8 | 3.4 |
| 1,2,3,4,7,8-hexaCDF | 0.2 | 0.1 | 1.3 | 0.2 | 0.9 |
| 1,2,3,6,7,8-hexaCDF | 0.2 | 0.08 | 0.6 | 0.2 | 0.5 |
| 1,2,3,7,8,9-hexaCDF | < 0.06 | < 0.03 | < 0.1 | < 0.06 | < 0.03 |
| 2,3,4,6,7,8-hexaCDF | 0.1 | 0.06 | 0.8 | 0.2 | 0.6 |
| 1,2,3,4,6,7,8-heptaCDF | 0.2 | 0.08 | 1.1 | 0.08 | 0.5 |
| 1,2,3,4,6,8,9-heptaCDF | < 0.2 | < 0.05 | < 0.5 | < 0.2 | < 0.06 |
| octaCDF | < 0.7 | 0.6 | 1.7 | < 0.2 | < 0.6 |
| 2,3,7,8-tetraCDD | 1.4 | 0.3 | 1.4 | 2.8 | 0.7 |
| 1,2,3,7,8-pentaCDD | 2.9 | 0.7 | 2.4 | 4.2 | 2.3 |
| 1,2,3,4,7,8-hexaCDD | 0.3 | 0.2 | 0.5 | 0.1 | 0.5 |
| 1,2,3,6,7,8-hexaCDD | 1.2 | 0.4 | 2.3 | 1.1 | 2.0 |
| 1,2,3,7,8,9-hexaCDD | 0.3 | 0.2 | 0.7 | 0.1 | 0.5 |
| 1,2,3,4,6,7,9-heptaCDD | 0.2 | 0.07 | 0.2 | 0.8 | 0.2 |
| 1,2,3,4,6,7,8-heptaCDD | 0.4 | 0.4 | 0.8 | 0.6 | 1.2 |
| octaCDD | 1.0 | 0.4 | 0.8 | 0.7 | 1.2 |
| PCDFs | 23.7 | 7.62 | 31.1 | 23.3 | 23.3 |
| PCDDs | 7.7 | 2.47 | 9.2 | 13.4 | 8.4 |
| PCDD + PCDF (total) | 31.4 | 10.09 | 40.3 | 36.7 | 31.7 |
| B. Bioassay (2,3,7,8-TCDD Equiv.) | | | | | |
| AHH | 89 | 51 | 44 | 76 ± 23 | 112 |
| EROD | 36 | ND | 22 | 37 ± 8 | 31 |

**Table 9.** PCDDs and PCDFs in Lake Ontario fish – GC-MS and bioassay results (parts per trillion)

| | Sample Code | | | | |
|---|---|---|---|---|---|
| | 84-10-106 | 84-10-107 | 84-10-108 | 84-10-109 | 84-10-110 |
| A. Congener | | | | | |
| 2,3,7,8-tetraCDF | 13 | 19 | 30 | 20 | 21 |
| 1,2,3,7,8-pentaCDF | 3.2 | 4.5 | 5.9 | 4.6 | 5.5 |
| 2,3,4,7,8-pentaCDF | 17 | 26 | 20 | 20 | 18 |
| 1,2,3,4,7,8-hexaCDF | 5.5 | 12 | 21 | 12 | 13 |
| 1,2,3,6,7,8-hexaCDF | 1.0 | 1.9 | 3.0 | 1.7 | 1.9 |
| 1,2,3,7,8,9-hexaCDF | < 0.2 | < 0.09 | < 0.05 | < 0.09 | < 0.06 |
| 2,3,4,6,7,8-hexaCDF | 0.6 | 2.3 | 0.7 | 1.3 | 0.9 |
| 1,2,3,4,6,7,8-heptaCDF | 0.4 | 0.8 | 1.5 | 1.0 | 0.7 |
| 1,2,3,4,6,8,9-heptaCDF | < 0.1 | < 0.1 | < 0.2 | < 0.06 | < 0.1 |
| octaCDF | < 0.3 | < 1.8 | < 0.8 | < 1.2 | < 0.2 |
| 2,3,7,8-tetraCDD | 37 | 42 | 45 | 43 | 36 |
| 1,2,3,7,8-pentaCDD | 8.0 | 10 | 8.2 | 7.4 | 8.6 |
| 1,2,3,4,7,8-hexaCDD | < 0.3 | 0.4 | 0.7 | 0.4 | 0.5 |
| 1,2,3,6,7,8-hexaCDD | 3.4 | 4.0 | 6.3 | 4.0 | 4.5 |
| 1,2,3,7,8,9-hexaCDD | < 0.6 | 0.6 | 0.5 | 0.3 | 0.3 |
| 1,2,3,4,6,7,9-heptaCDD | 0.4 | 0.1 | < 0.3 | 0.2 | < 0.8 |
| 1,2,3,4,6,7,8-heptaCDD | 1.6 | 0.5 | 0.7 | 0.6 | 1.7 |
| octaCDD | 0.9 | 1.8 | 1.0 | 0.8 | 0.9 |
| PCDFs | 37.7 | 66.2 | 82.1 | 60.6 | 61.0 |
| PCDDs | 51.3 | 59.4 | 62.4 | 56.7 | 54.5 |
| PCDD + PCDF (total) | 89.0 | 125.6 | 144.5 | 117.3 | 115.5 |
| B. Bioassay (2,3,7,8-TCDD Equiv.) | | | | | |
| AHH | 378 | 410 | 405 | 501 | 308 |
| EROD | 137 | 310 | 273 | 393 | 302 |

1,2,3,6,7,8-hexaCDD. These compounds were also the major PCDD and PCDF congeners in the Lake Ontario fish extracts; however, in these extracts the total PCDD and PCDF levels were comparable. This was primarily due to the unusually high concentrations of 2,3,7,8-TCDD. These results suggest that Lake Ontario contains additional point sources of 2,3,7,8-TCDD which are probably associated with 2,4,5-trichlorophenol-derived chemical wastes containing high levels of 2,3,7,8-TCDD. Figure 4 compares the relative levels of the total concentrations of PCDDs and PCDFs found in the fish samples from the Great Lakes and the "2,3,7,8-TCDD equivalents" which were estimated using 2 sensitive bioassays, namely the induction of aryl hydrocarbon hydroxylase (AHH) and ethoxyresorufin *O*-deethylase (EROD) activities in rat hepatoma H-4-II E cells in culture [89–94]. For most of these samples, there was less than a 2-fold difference in the bioassay-estimated 2,3,7,8-TCDD equivalents and the total PCDDs plus PCDFs as determined by GC-MS. The bioassay-derived values were significantly ( > 2-fold) higher for the Lake Erie and Lake Ontario fish extracts. This difference may be due to several factors including synergistic interactive effects of the congeners in the bioassay induction response or the presence of "bioassay-active" components that are not detected by GC-MS

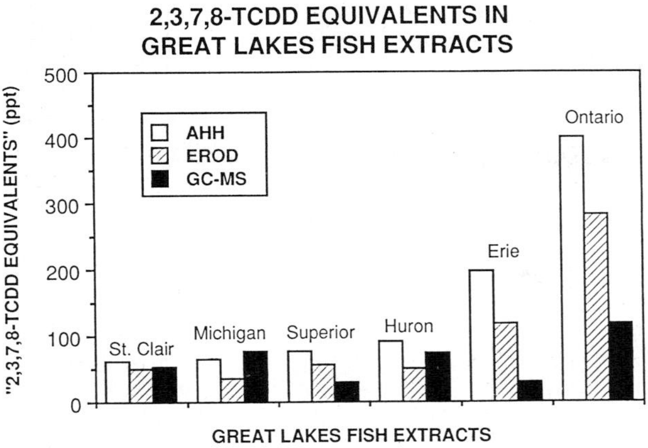


**Fig. 4.** Comparative PCDD and PCDF levels in Great Lakes fish extracts: GC-MS analysis and bioassay results using the induction of AHH and EROD activities in rat hepatoma H-4-II E cells in culture

analysis. Some of the compounds could be bioactive bromo/chloro PCDDs and PCDFs which have also been identified as by products in brominated industrial compounds, chemical waste combustion and in car exhausts [95–101].

Residues of PCDFs and PCDDs have been extensively identified in human adipose tissue, blood and milk in individuals in many different countries [21, 102, 103]. In most adipose tissue samples the overall PCDD levels are higher than the concentration of PCDFs, and octaCDD is the dominant congener in most samples. Tables 10 and 11 summarize the PCDD and PCDF levels in human adipose tissue and milk samples from several countries. These data illustrate that only the toxic 2,3,7,8-substituted congeners are present in these samples; however, it is also evident that the bioconcentration potential of individual PCDD and PCDF congeners is very different in humans (Tables 10 and 11) and fish (Tables 5 and 6). In fish, the dominant PCDFs are 2,3,7,8-TCDF, 2,3,4,7,8-pentaCDF and 1,2,3,4,7,8-hexaCDF. In humans, there are also high levels of the penta- and hexaCDF congeners; however, relatively low concentrations of 2,3,7,8-TCDF were detected. OctaCDD is the major congener identified in humans; however, this higher chlorinated compound is a minor constituent of the total PCDDs in fish. In addition, the higher chlorinated PCDDs are more readily bioconcentrated in humans than in fish.

Based on the distribution of PCDDs and PCDFs in human tissue, it is not possible to assign specific sources for these compounds although it has been suggested that combustion-derived processes are major contributors to the bioavailable PCDDs and PCDFs [21]. However, it should also be noted that PCDDs and PCDFs have been reported in a variety of food products including fish, domestic food animals, cow's milk, butter, chickens and eggs [104].

**Table 10.** Levels of PCDFs in human tissues

| Congener | Sample Source and Mean Concentrations (ppt on fat basis) | | |
|---|---|---|---|
| A. Adipose Tissue | Japan (13)[a] | Sweden (31)[b] | FRG[c] |
| 2,3,7,8-tetraCDF | 9 | 3.9 | 0.9 |
| 2,3,4,7,8-pentaCDF | 25 | 54 | 44 |
| 1,2,3,4,7,8-hexaCDF | 15 | 6 | 10.0 |
| 1,2,3,6,7,8-hexaCDF | 14 | 5 | 6.7 |
| 2,3,4,6,7,8-hexaCDF | 8 | 2 | 3.8 |
| 1,2,3,4,6,7,8-heptaCDF | nr | 11 | 19.5 |
| octaCDF | nr | 4 | < 1 |
| B. Milk | FRG[d] | Sweden (4)[e] | Netherlands[c] |
| 2,3,7,8-tetraCDF | 2.6 | 4.2 | 2.8 |
| 2,3,4,7,8-pentaCDF | 22.9 | 21.3 | 79 |
| 1,2,3,4,7,8-hexaCDF | 8.2 | 4.7 | 8.9 |
| 1,2,3,6,7,8-hexaCDF | 6.6 | 3.4 | 10.3 |
| 2,3,4,6,7,8-hexaCDF | 3.3 | 1.4 | 6.4 |
| 1,2,3,4,6,7,8-heptaCDF | 6.4 | 7.4 | 39 |
| octaCDF | 22.8 | 3.2 | ND |

[a] Ref. [117]; [b] Ref. [118]; [c] Ref. [21]; [d] Ref. [119]; [e] Ref. [68]

**Table 11.** Level of PCDDs in human tissues

| Congener | Sample Source and Mean Concentrations (ppt on fat basis) | | |
|---|---|---|---|
| A. Adipose Tissue | Japan (13)[a] | Sweden (31)[b] | FRG (4)[c] |
| 2,3,7,8-tetraCDD | 9 | 3 | 150 |
| 1,2,3,7,8-pentaCDD | 15 | 10 | 19.2 |
| 1,2,3,6,7,8-hexaCDD | 70 | 15 | 77 |
| 1,2,3,7,8,9-hexaCDD | 12 | 4 | 9.4 |
| 1,2,3,4,6,7,8-heptaCDD | 77 | 97 | 56 |
| octaCDD | 230 | 414 | 267 |
| B. Milk | FRG[d] | Sweden (4)[e] | Netherlands[c] |
| 2,3,7,8-tetraCDD | < 5 | 0.6 | 9.7 |
| 1,2,3,7,8-pentaCDD | 10.7 | 6.5 | 44 |
| 1,2,3,4,7,8-hexaCDD | 8.1 | 2.5 | 25 |
| 1,2,3,6,7,8-hexaCDD | 32.7 | 19 | 251 |
| 1,2,3,7,8,9-hexaCDD | 6.4 | 6.3 | 23 |
| 1,2,3,4,6,7,8-heptaCDD | 49.9 | 59.5 | 130 |
| octaCDD | 181.2 | 302 | 744 |

[a] Ref. [114]; [b] Ref. [115]; [c] Ref. [20]; [d] Ref. [116]; [e] Ref. [67]

### *3.4 Yusho and Yu-Cheng Incidents*

In 1968, a mass food poisoning was reported in the Fukuoka and Nagasaki prefectures in southwestern Japan and approximately 1600 individuals suffered a broad spectrum of toxic effects after consuming rice oil contaminated with a commercial PCB industrial fluid, Kanechlor 400 [105–107]. The 1979 Yu-Cheng poisoning of > 1900 individuals in Taichung and Changhwa in central Taiwan was also due to the consumption of PCB-contaminated rice oil [108, 109]. It was apparent that there were major differences in the toxic effects of PCB contaminated rice oil and "normal" industrial PCBs on their respective exposed human populations. The severe acute and chronic effects observed in Yusho victims consuming the contaminated rice oil were not observed in the occupationally-exposed population. Moreover, a close inspection of serum PCB levels in Yusho/Yu-Cheng patients and industrial workers exposed to PCBs were comparable. For example, the mean PCB blood levels from Yu-Cheng victims taken a short time after the accident varied from 39–191.7 ppb; whereas, the PCB serum levels in occupationally-exposed workers can be much higher [110]. However, it was also apparent that the PCDF levels in the industrial PCBs which leaked into the rice oil were much higher than the corresponding PCDF concentrations in commercial PCBs [8, 12, 108, 109, 111, 112]. Subsequent studies have demonstrated that the more highly toxic PCDFs were the major etiologic agents in the Yusho and Yu-Cheng poisonings [113–115]. Analysis of the adipose tissue and liver for PCDFs in 4 Yusho patients showed that relatively high levels of several 2,3,7,8-substituted PCDFs were present in these tissues [15]. The concentrations of the total PCDFs in liver (0.16 to 17.6 ppb) and adipose tissue (0.7 to 9.3 ppb) were significantly higher than those observed in environmentally exposed individuals (see Tables 10 and 11).

This chapter briefly summarizes the major sources of PCDDs and PCDFs and their levels in different components of the global ecosystem. No doubt, new sources of these compounds will be found in future studies and this will facilitate estimations of the relative inputs of PCDDs and PCDFs from different sources. The remaining chapters in this book will describe the toxicology, mechanism of action and human health effects of PCDDs and PCDFs and utilize this information to assess the potential human health hazards and risks associated with exposure to these compounds.

*Acknowledgements.* The financial assistance of the National Institutes of Health (P42-ES04917) and the Texas Agricultural Experiment Station to S.S. are gratefully acknowledged. O.H. wishes to thank Dr. H. Fiedler for help with the literature.

## 4 References

1. Heindl A, Hutzinger O (1986) Chemosphere 15: 653
2. Herrick EC, Goldfarb AS, Fong CV, Konz J, Walker P Chlorophenols by chlorination of phenol. In: Hazards associated with organic chemical manufacturing. MITRE Technical Report, MTR-78 W00364-05

3. Hagenmaier H, Brunner H (1987) Chemosphere 20: 2425
4. Christmann W, Kloppel KD, Partscht H, Rotard W (1989) Chemosphere 18: 861
5. Vos JG, Koeman JH, van der Maas HL, ten Noever de Brauw MC, de Vos RH (1970) Food Cosmet. Toxicol. 8: 625
6. Roach JAG, Pomerantz IH (1974) Bull. Environ. Contam. Toxicol. 12: 338
7. Bowes GW, Mulvihill MJ, Decamp MR, Kende AS (1975) J. Agric. Food Chem. 23: 1222
8. Morita M, Nakagawa J, Akiyama K, Mimura S, Isono N (1977) Bull. Environ. Contam. Toxicol. 18: 67
9. Rappe C, Gara A, Buser HR, Bosshardt H-P (1977) Chemosphere 6: 231
10. Rappe C, Nygren M, Buser HR, Masuda Y, Kuroki H, Chen PH (1983) Environ. Sci. Res. 26: 241
11. Nagayama J, Kuratsune M, Masuda Y (1976) Bull. Environ. Contam. Toxicol. 15: 9
12. Buser HR, Rappe C, Gara A (1978) Chemosphere 7: 439
13. Wakimoto T, Kannan N, Ono M, Tatsukawa R, Masuda Y (1988) Chemosphere 17: 743
14. Brown JF, Carnahan JC, Dorn SB, Groves JT, Ligon WV, May RJ, Wagner RE, Hamilton SB (1988) Chemosphere 17: 1697
15. Masuda Y, Kuroki H, Haraguchi K, Nagayama J (1986) Chemosphere 15: 1621
16. Villanueva EC, Jennings RW, Burse VW, Kimbrough RD (1974) J. Agric. Food Chem. 22: 916
17. Scholz B, Engler M (1987) Chemosphere 16: 1829
18. Knuutinen J, Salovaara J, Tarhanen J, Passivirta J, Virkki L, Lahtipera M, Humppi T, Laitinen R, Kantolahti E (1983) Chemosphere 12: 511
19. Leuenberger C, Coney R, Graydon JW, Molnar-Kubica E, Giger W (1983) Chimia 37: 345
20. Paasivirta J, Tarhanen J, Soikkeli J (1986) Chemosphere 15: 1429
21. Rappe C, Andersson R, Bergqvist PA, Brohede C, Hansson M, Kjeller LO, Lindstrom G, Marklund S, Nygren M, Swanson SE, Tysklind M, Wiberg K (1987) Chemosphere 16: 1603
22. Kuehl DW, Butterworth BC, DeVita WM, Sauer CP (1987) Biomed. Environ. Mass Spectrom. 14: 443
23. Swanson SE, Rappe C, Malmstrom J, Kringstad KP (1988) Chemosphere 17: 681
24. Clement RE, Tashiro C, Suter S, Reiner E, Hollinger D (1989) Chemosphere 18: 1189
25. Amendola G, Barna D, Blosser R, LaFleur L, McBride A, Thomas F, Tiernan T, Whittemore R (1989) Chemosphere 18: 1181
26. Beck H, Eckart K, Mathar W, Wittkowski R (1989) Chemosphere 17: 51–75
27. Kitunen V-H, Salkinoja-Salonen MS (1988) Occurrence of PCDDs and PCDFs in pulp and board products, SOUP05 presented at DIOXIN '88, Umea, Sweden, August 21–26, 1988
28. Olie K, Vermeulen P, Hutzinger O (1977) Chemosphere 6: 455
29. Crummett WB, Townsend DI (1984) Chemosphere 13: 777
30. Tiernan TO, Taylor ML, Garrett JH, VanNess GF, Solch JG, Deis DA, Wagel DG (1983) Chemosphere 12: 595
31. Hutzinger O, Blumich MJ, v.d. Berg M, Olie K (1985) Chemosphere 14: 581
32. Clement RE, Tosine HM, Osborne J, Ozvacic V, Wong G (1988) Biomed. Environ. Mass Spectrom. 17: 81
33. Benfenati E, Gizzi F, Reginato R, Fanelli R, Lodi M, Tagliaferri R (1983) Chemosphere 12: 1151
34. Nottrodt IA, Ballschmiter K (1986) Chemosphere 15: 1225
35. Thoma H (1987) PCDD/F-Konzentrationen in Kaminaschen bei Hausfeuerung. In: VDI Berichte 634, VDI Verlag, Dusseldorf. p 53
36. Hutzinger O, Fiedler H (1989) Chemosphere 18: 23
37. Thoma H (1988) Chemosphere 17: 1369
38. Thoma H (1988) PCDD/F-concentrations in chimney soot from house heating systems, Dioxin '88, 8th International Symposium on Chlorinated Dioxins and Related Compounds, 21–28 August 1988, Umea, Sweden
39. Czuczwa JM, Hites RA (1984) Environ. Sci. Technol. 18: 444
40. Czuczwa JM, Hites RA (1985) Dioxins and dibenzofurans in air, soil and water. In: Kamrin MA, Rodgers PW (eds) Dioxins in the environment Hemisphere, Washington, p B5–99
41. Czuczwa JM, Hites RA (1986) Environ. Sci. Technol. 20: 195
42. Czuczwa JM, McVeety BD, Hites RA (1984) Science 226: 568
43. Czuczwa JM, Niessen F, Hites RA (1985) Chemosphere 14: 1175
44. Chiu C, Thomas RS, Lockwood J, Li K, Halman R, Lao RCC (1983) Chemosphere 12: 607
45. Clement RE, Tosine HM, Ali B (1985) Chemosphere 14: 815

46. Clement RE, Tosine HM, Osborne J, Ozvacic V, Wong G, Thorndyke S (1987) Chemosphere 16: 1895
47. Buser HR, Rappe C (1979) Chemosphere 8: 157
48. Buser HR, Bosshardt H-P, Rappe C (1978) Chemosphere 7: 109
49. Buser HR, Bosshardt H-P, Rappe C (1978) Chemosphere 7: 419
50. des Rosiers PE (1987) Chemosphere 16: 1881
51. Schecter A (1987) Chemosphere 16: 2147
52. Elo O, Vuojolahti P, Janhunen H, Rantanen J (1985) Environ. Health Perspect. 60: 315
53. Rappe C, Marklund S, Bergqvist PA, Hansson M (1983) Polychlorinated dibenzo-*p*-dioxins, dibenzofurans and other polynuclear aromatics formed during incineration and polychlorinated biphenyl fires. In: Choudhary G, Keith LH, Rappe C (eds) Chlorinated dioxins and dibenzofurans in the total environment, Butterworth, Boston, MA, p 99
54. Rappe C, Marklund S, Buser HR, Bosshardt HP (1978) Chemosphere 7: 269
55. Zoller W, Ballschmiter K (1986) Fres. Zeit. Anal. Chem. 323: 19
56. Ahling B, Lindskog A, Jansson B, Sundstrom G (1977) Chemosphere 8: 461
57. Ahling B, Bjorseth A, Lune G (1978) Chemosphere 8: 799
58. Marklund S, Kjeller L-O, Hansson M, Tysklind M, Rappe C, Ryan C, Collazo H, Dougherly R (1986) Determination of PCDDs and PCDFs in incineration samples and pyrolytic products. In: Rappe C, Choudhary G, Keith LH (eds) Chlorinated dioxins and dibenzofurans in perspective, Lewis, Chelsea. p 79
59. Buser HR (1979) Chemosphere 8: 415
60. Muller MD, Buser H-R (1986) Environ. Sci. Technol. 20: 1151
61. Marklund S, Rappe C, Tysklind M, Edgebaсk K-E (1987) Chemosphere 16: 29
62. Choudhry GG, Olie K, Hutzinger O (1982) Mechanisms in the thermal formation of chlorinated compounds including polychlorinated dibenzo-*p*-dioxins. In: Hutzinger O, Frei RW, Merian E, Pocchiari F (eds) Chlorinated dioxins and related compounds. Impact on the environment, Pergamon, Oxford, p 275
63. Choudhry GG, Broecke JAVD, Hutzinger O (1983) Chemosphere 12: 487
64. Janssens JJ, Schepens PJC (1988) Biomed. Environ. Mass Spectrom. 16: 179
65. Ballschmiter K, Braunmiller I, Niemczyk R, Swerev M (1988) Chemosphere 17: 995
66. Karašek FW, Dickson LC (1987) Science 237: 754
67. Eklund G, Pederson JR, Stromberg B (1988) Chemosphere 17: 575
68. Rappe C, Nygren M, Lindstrom G, Hansson M (1986) Chemosphere 15: 1635
69. Rappe C, Kjeller L-O, Bruckmann P, Hackhe K-H (1988) Chemosphere 17: 3
70. Nakano T, Tsuji M, Okuno T (1987) Chemosphere 16: 1781
71. Swerev M, Ballschmiter K (1989) Chemosphere 18: 609
72. Reischl A, Reissinger M, Thoma H, Hutzinger O (1989) Chemosphere 18: 561
73. Rappe C, Kjeller L-O (1987) Chemosphere 16: 1775
74. Evers EGH, Ree KCM, Olie K (1988) Chemosphere 17: 2271
75. Weerasinghe NCA, Gross ML, Lisk DJ (1985) Chemosphere 14: 557
76. Heida H, Olie K (1985) Chemosphere 14: 919
77. Harless RL, Oswald EO, Lewis RG, DuPuy AE, McDaniel DD, Tai H (1982) Chemosphere 11: 193
78. O'Keefe P, Meyer C, Hilker D, Aldous K, Jelus-Tyror B, Dillon K, Donnely R (1983) Chemosphere 12: 325
79. Rappe C, Buser HR, Stalling DL, Smith LM, Dougherty RC (1981) Nature. 292: 524–526
80. Stalling DL, Smith LM, Petty JD, Hogan JW, Johnson JL, Rappe C, Buser HR (1983) Residues of polychlorinated dibenzo-*p*-dioxins and dibenzofurans in Laurentian Great Lakes fish. In: Tucker RE, Young AL, Gray AP (eds) Human and environmental risks of chlorinated dioxins and related compounds, Plenum, p 221
81. Stalling DL, Norstrom RJ, Smith LM, Simon M (1985) Chemosphere 14: 627
82. Ryan JJ, Lau PY, Pilon JC, Lewis D (1983) 2,3,7,8-Tetrachlorodibenzo-*p*-dioxin and 2,3,7,8-tetrachlorodibenzofuran residues in Great Lakes commercial and sport fish. In: Choudhary G, Keith LH, Rappe C (eds) Chlorinated dioxins and dibenzofurans in the total environment, Butterworth, Boston, MA, p 87
83. van den Berg M, Blank F, Heeremans C, Wagenaar H, Olie K (1987) Arch. Environ. Contam. Toxicol. 16: 149
84. Oehme M, Furst P, Kruger C, Meemken HA, Groebel W (1988) Chemosphere 17: 1291
85. Miyata H, Takayama K, Ogaki J, Kashimoto T, Fukushima S (1987) Bull. Environ. Contam. Tox. 39: 877

86. Miyata H, Takayama K, Ogaki J, Mimura M, Kashimoto T (1988) Toxicol. Environ. Chem. 17: 91
87. Rappe C, Bergqvist PA, Kjeller LO (1989) Chemosphere 18: 651
88. Zacharewski T, Safe L, Safe S, Chittim B, DeVault D, Wiberg K, Bergqvist PA, Rappe C (1989) Environ. Sci. Technol. 23: 730
89. Bradlaw JA, Casterline JL Jr (1979) J. Assoc. Off. Anal. Chem. 62: 904
90. Mason G, Sawyer T, Keys B, Bandiera S, Romkes M, Piskorska-Pliszczynska J, Zmudzka B, Safe S (1985) Toxicol. 37: 1
91. Mason G, Farrell K, Keys B, Piskorska-Pliszczynska J, Safe L, Safe S (1986) Toxicol. 41: 21
92. Mason C, Zacharewski T, Denomme MA, Safe L, Safe S (1987) Toxicol. 44: 245
93. Safe SH (1986) Annu. Rev. Pharmacol. Toxicol. 26: 371
94. Safe S (1987) Chemosphere 16: 791
95. Buser HR (1986) Anal. Chem. 58: 2913
96. Buser HR (1987) Chemosphere 16: 713
97. Thoma H, Rist S, Hausschulz G, Hutzinger O (1986) Chemosphere 15: 649
98. Hileman F, Wehler J, Wending J, Orth R, Ritchie C, McKenzie D (1989) Chemosphere 18: 217
99. Schafer W, Ballschmiter K (1986) Chemosphere 15: 755
100. Haglund P, Egeback K-E, Jansson B (1988) Chemosphere 17: 2129
101. Sovocool GW, Donnelly JR, Munslow WD, Vonnahme TL, Nunn NJ, Tondeur Y, Mitchum RK (1989) Chemosphere 18: 193
102. Schecter A, Ryan JJ, Constable JD (1989) Chemosphere 18: 975
103. Beck H, Eckart K, Mathar W, Wittkowski R (1989) Chemosphere 18: 1063
104. Beck H, Eckart K, Mathar W, Wittkowski R (1989) Chemosphere 18: 417
105. Kuratsune M, Yoshimura T, Matsuzaka J, Yamaguchi A, (1972) Environ. Health Persp. 1: 119
106. Kuratsune M (1975) National Conference on Polychlorinated Biphenyls. Chicago, November 19–21, 1975. EPA/6-75-004
107. Kuratsune M (1980) Yusho. In: Kimbrough RD (ed) Halogenated biphenyls, terphenyls, naphthalenes, dibenzodioxins and related products, Elsevier, Amsterdam, p 287
108. Chen PH, Chang KT, Lu YD (1981) Bull. Environ. Contam. Toxicol. 26: 489
109. Chen PH, Wong CK, Rappe C, Nygren M (1985) Environ. Health Perspect. 59: 59
110. Hsu ST, Ma CI, Kwo-Hsiang S, Wu SS, Hsu NH-M, Yeh CC, Wu SB (1985) Environ. Health Perspect. 59: 5
111. Masuda Y, Kuroki H, Yamaryo T, Haraguchi K, Kuratsune M, Hsu ST (1982) Chemosphere 11: 199
112. Miyata H, Kashimoto T, Kunita N (1977) J. Food Hyg. Soc. 18: 260
113. Kunita N, Kashimoto T, Miyata H, Fukushima S, Hall S, Obana H (1984) Amer. J. Ind. Med. 5: 45
114. Kunita N, Hori S, Obana H, Otake T, Nishimura H, Kashimoto T, Ikegami N (1985) Environ. Health Perspect. 59: 79
115. Bandiera S, Farrell K, Mason G, Kelley M, Romkes M, Bannister R, Safe S (1984) Chemosphere 13: 507
116. Norstrom A, Rappe C, Lindahl R, Buser HR (1979) Scand. J. Work Environ. Health 5: 375
117. Ono M, Wakimoto T, Tatsukawa R, Masuda Y (1986) Chemosphere 86: 16
118. Nygren M, Hansson M, Sjostrom M, Rappe C, Kahn P, Gochfeld M, Velez H, Ghent-Guenther T, Wilson WP (1988) Chemosphere 17: 1663
119. Furst P, Meemken H-A, Kruger C, Groebel W (1987) Chemosphere 16: 1983

# 2 UAREP-Report on Health Aspects of Polychlorinated Dibenzo-*p*-dioxins (PCDDS) and Polychlorinated Dibenzofurans (PCDFs)

**Ad Hoc Panel**

**Health Aspects of Polychlorinated Dibenzo-*p*-dioxins (PCDDs) and Polychlorinated Dibenzofurans (PCDFs)**

*Chairman*

Robert E. Stowell, M.D., Ph.D., University of California, Davis

*Epidemiology*

Leonard T. Kurland, M.D., Dr. P.H., Mayo Clinic
Sidney Shindell, M.D., LL.B., Medical College of Wisconsin
Allan H. Smith, M.D., Ph.D., University of California, Berkeley

*Mechanism*

Thomas Gasiewicz, Ph.D., University of Rochester
Stephen H. Safe, Ph.D., Texas A&M University
James P. Whitlock, Jr., M.D., Stanford University

*Risk Characterization*

David R. Brown, Sc.D., Connecticut Department of Health Services
Ian C. Munro, Ph.D., F.R.C. Path., Canadian Center for Toxicology
Carl O. Schulz, Ph.D., University of South Carolina

## Preface

This chapter reviews the health aspects of the environmental distribution of polychlorinated dibenzodioxins (PCDDs) and polychlorinated dibenzofurans (PCDFs). It is adapted from the Final Report of the Ad Hoc Panel on the Health Aspects of PCDDs and PCDFs convened under the auspices of the Board of Directors of Universities Associated for Research and Education in Pathology, Inc. (UAREP). This Panel was tasked to examine the scientific literature on PCDDs/PCDFs and respond to the following questions on the current knowledge and understanding of the health effects of environmental exposure to these closely related families of chemicals.

Q. (1) What do epidemiological studies tell us about the health effects of dioxins in humans?
Q. (2) What is the strength of evidence that dioxin's effects are receptor mediated?
Q. (3) If dioxin acts through a receptor binding mechanism, is it reasonable to assume that its effects are reversible and that a practical threshold exists?
Q. (4) Is there a scientifically supportable method for quantifying the human health hazard of exposure to dioxin?
  a) If so, what is the biological basis for interpreting the risk of low level exposure?
  b) If not, based on the available animal, human, and mechanistic data, how would the Panel characterize the hazard to humans of part-per-billion or part-per-trillion levels of dioxins in the environment?

The Panel reviewed the published literature on the epidemiology, mechanisms of effect, and elements of risk characterization for PCDDs and PCDFs, and three of its members prepared draft chapters on these topics. The members of the Panel formed working groups which revised and further developed the initial drafts to create a Draft Report acceptable to the Panel as a whole. The Final Report was reviewed and approved by a committee of the Board of Directors prior to its release in June, 1988.

The Panel endeavored to consider all relevant arguments and to express its interpretation of the scientific literature in the context of the weight of evidence, but specific findings and conclusions should not be construed to represent the personal opinion or convictions of individual Panel members.

The Chairman expresses his sincere appreciation to the authors of the draft chapters, Dr. Stephen Safe, Dr. Carl Schulz, and Dr. Allan Smith, and to the other members of the Panel for their dedicated efforts in the development of the basic report. The Panel is especially grateful to Thomas A. Hill, Project Director, for compiling and editing the report for inclusion in this monograph. The initial study was supported in part by an unrestricted grant from the American Paper Institute.

# 2.1 Epidemiology

**Epidemiology Working Group**[1]

**A.H. Smith, L.T. Kurland and S. Shindell**

1 Introduction . . . 27
2 Effects Definitely or Probably Caused by PCDDs/PCDFs . . . 29
2.1 Effects Attributable to PCDDs . . . 30
2.2 Effects Attributable to PCDFs . . . 30
2.3 Conclusions on Definite or Probable Effects . . . 31
3 Effects Possibly Caused by PCDDs/PCDFs . . . 31
3.1 Porphyria . . . 31
3.2 Changes in Blood Lipids and Indicators of Liver Function . . . 32
3.3 Gastrointestinal Ulcer . . . 33
3.4 Immune System Effects . . . 34
3.5 Neurotoxicity . . . 35
3.6 Conclusions on Possible Effects . . . 35
4 Effects on Reproduction . . . 36
4.1 Studies With Ecological Exposure Data . . . 36
4.2 Studies Identifying Exposed Individuals: Herbicide Spraying . . . 39
4.3 Studies Identifying Exposed Individuals: Chemical Manufacture . . . 39
4.4 Conclusions Concerning Reproductive Effects . . . 40
5 Carcinogenicity Studies in Humans . . . 40
5.1 Introduction . . . 40
5.2 Soft Tissue Sarcoma . . . 41
5.3 Malignant Lymphoma . . . 47
5.4 Cancer of Other Sites . . . 50
5.5 Interpretation of Human Cancer Findings . . . 52
6 Summary and Conclusions . . . 53
7 References . . . 54

## 1 Introduction

This chapter is a review of epidemiological studies of occupational and environmental exposures to polychlorinated dibenzo-*p*-dioxins (PCDDs) and polychlorinated dibenzofurans (PCDFs).

[1] Ad Hoc Panel on Health Aspects of Polychlorinated Dibenzo-*p*-dioxins and Polychlorinated Dibenzofurans, Universities Associated for Research and Education in Pathology, Inc., Bethesda, Maryland, USA

"The basic premise of epidemiology is that disease does not occur randomly but in patterns which reflect the operation of the underlying causes; ... knowledge of these patterns is not only of predictive value with respect to future disease occurrence, but also constitutes a major key to understanding causation ...." (Fox et al., 1970).

Epidemiological studies are generally observational in nature. In cohort studies, health outcomes are ascertained for a group of people with definable characteristics such as employment, and the incidence of disease is compared between those exposed and not exposed, or between persons exposed at different levels to a given agent. The relative risk (RR) (or rate ratio) estimate is the incidence rate among the exposed divided by the incidence rate among the non-exposed. Cohort studies, particularly occupational cohort studies, are very important in assessing possible risk of cancer from exposures in excess of those usually encountered by the general public, but large cohorts are required to detect increased risks for rare diseases. In case-control studies, the starting point is the identification of a group of people with a disease, followed by selection of a comparison (control) group without the disease. Past exposures are compared between the two groups, and an estimate of relative risk can be calculated. While case-control comparisons are of value for studying rare diseases, they present two important problems. First, the exposure must be sufficiently frequent in the population in order to achieve sufficient statistical power. Second, case-control studies are subject to various types of bias which are not present in cohort studies. Selection bias may result from the choice of inappropriate controls, and the ascertainment of past exposures among the cases or controls may be influenced by investigators (interviewer bias) or by the study subjects themselves (recall bias).

There is a variety of other types of epidemiological studies. Ecological studies compare disease frequency among population groups located in areas which differ in the extent of exposure. Other studies look for changes of disease frequency over time compared to trends in exposure. These studies have several limitations when it comes to causal inference, because many factors vary with location and time.

Criteria for causal inference in epidemiology must account for the observational (rather than experimental) nature of the data collected. Even when findings are unlikely to be due to chance, known or unknown sources of bias may be present. The observations in any one study could be due to misclassification of exposure or disease leading to biased estimates of relative risk, or to confounding associations between the exposure under study and other factors which are causally related to the disease. Epidemiologists, therefore, give considerable weight to the consistency of findings in multiple studies of potential effects of an exposure. Bias may produce spurious positive associations, negative associations, or absence of associations. Epidemiologists do not expect absolute consistency from study to study, but consistency plays a very important role in causal inference. If one study finds a substantial increase in risks for a particular exposure level, then most studies involving exposures at the same level or higher should find elevation in relative risk estimates if the initial findings reflected a causal association.

The strength of associations found is also important in epidemiological causal inference. A group of studies finding relative risk estimates around 1.5, for example, even if unlikely to be due to chance, is much less convincing than a group of studies reporting relative risks of around 5. At the same time, it should be recognized that some studies might be expected to report high relative risks attributable to chance; therefore, strength of association should not be interpreted separately from consideration of statistical significance. The presence of a dose-response relationship also strengthens causal inference from epidemiological studies. True causal associations between chemical exposures and toxic effects should normally manifest a positive relationship between the extent of exposure and the incidence of effects; however, a variety of types of bias, especially confounding, may lead to spurious, monotonic dose-response relationships.

Epidemiological studies must also be considered in the light of biological plausibility, albeit cautiously, because the first discovery of an association between exposure and disease frequently lacks sufficient biological information to assess its plausibility. An increased tumor incidence among animals exposed to a particular chemical in laboratory bioassay studies increases the biological plausibility for similar findings in epidemiological studies. The review which follows considers various potential effects of PCDDs and PCDFs in humans in the light of these principles. Although the PCDDs and PCDFs may have been the most potent toxins involved in these studies, none of the exposures were limited solely to these two classes of compounds. In most cases the PCDDs or PCDFs were minor contaminants of complex mixtures of biologically active halogenated aryl hydrocarbons. Concurrent exposures most frequently included polyhalogenated phenoxyacetic and phenoxybutyric acids, polychlorinated phenols, and polyhalogenated biphenyls and quaterphenyls. For this reason all epidemiological evidence must be interpreted in the context of multiple and potentially interactive exposures.

The review first considers outcomes definitely or probably caused by PCDDs/PCDFs for which there is strong evidence for a causal relationship. The second section reviews effects which are possibly due to PCDDs/PCDFs, but the evidence is more limited. The third section presents outcomes for which there is consistent evidence suggesting no effect. Finally, consideration is given to the inconsistent associations surrounding human cancer studies.

## 2 Effects Definitely or Probably Caused by PCDDs/PCDFs

There are a variety of sources of data concerning human health effects of PCDDs, while information concerning the effects of PCDFs is confined to the Yusho and Yu-Cheng episodes.

## 2.1 *Effects Attributable to PCDDs*

Chloracne is a well-known human effect of exposure to PCDDs, PCDFs, and some other halogenated aromatic compounds. The evidence that chloracne can be caused by exposure to PCDDs and PCDFs is convincing, and will not be presented in detail here. Chloracne has been reported from workplace exposure after industrial accidents involving the release of 2,3,7,8-TCDD, from general population exposure after the industrial accident in Seveso, Italy, and in association with PCDF and PCB exposure in the Yusho and Yu-Cheng episodes involving contaminated rice oil (Caramaschi et al., 1981; Crow, 1983; Reggiani, 1983; Suskind, 1985; Tindall, 1985).

Chloracne can be confidently linked to PCDD/PCDF exposure for several reasons. It has been consistently reported to occur soon after heavy exposure, and in some industrial accidents involving release of 2,3,7,8-TCDD, almost all exposed workers developed chloracne within a few weeks of exposure. Furthermore, the heavier the exposure, the more severe the resulting chloracne. Thus, the association between chloracne and exposure to PCDDs/PCDFs is a strong one. It is consistent from study to study; there is a positive dose-response association; and, it is biologically plausible on the basis of animal experimentation.

## 2.2 *Effects Attributable to PCDFs*

Yusho and Yu-Cheng refer to two incidents of mass food poisoning caused by consumption of rice oils contaminated with polychlorinated biphenyls (PCBs). The PCBs were used as heat exchanger fluids during manufacture of the oil. Yusho or "oil disease" occurred in 1968 in western Japan when more than 1600 people were affected. Yu-Cheng (the Chinese equivalent of Yusho) occurred in 1979 in central Taiwan when approximately 2000 individuals were affected. Both of these affected populations have been extensively studied and described since the times of these incidents. These incidents and the ensuing health effects were reviewed in the proceedings of the Japan-United States Joint Seminar on Toxicity of Chlorinated Biphenyls, Dibenzofurans, Dibenzodioxins and Related Compounds (Japan-United States, 1985).

The characteristic manifestations reported by Yusho and Yu-Cheng victims (Kuratsune et al., 1972; Kunita et al., 1985; Chen and Hsu, 1987; Kashimoto and Miyata, 1987) included: 1) dermal signs: acne on the face, back, and external genitalia; dark pigmentation on face, nails, and gingivae; facial edema, especially on the eyelids; hyperkeratosis of the trunk and external genitalia; and, hyperhidrosis; 2) ocular signs: dark pigmentation of the conjunctiva; swelling of the Meibomian glands; and, hypersecretion of a cheese-like discharge from the eyes; 3) respiratory signs: chronic bronchitis with a persistent cough and sputum; 4) neurological signs and symptoms: headache; numbness of the limbs; a decline in vision; mild peripheral neuropathy; and, decreased nerve conduction velocities; 5) other signs and symptoms: irregular menstrual cycles; general fatigue; anorexia; hepatic damage; immunologic impairment; and, biochemical alterations.

Many of these symptoms and signs have also been reported in humans following exposure to halogenated aromatic compounds including dioxins, furans, and biphenyls, but only the ocular signs (swelling and hypersecretion of the Meibomian glands) and chloracne have been clearly connected to the exposure.

Initially, the signs and symptoms exhibited by the victims of the Yusho and Yu-Cheng incidents were attributed to PCBs (Kuratsune et al., 1972). Subsequent analyses of the samples of the contaminated rice oils indicated that the PCBs were also contaminated with PCDFs and polychlorinated quaterphenyls (Kashimoto et al., 1981; Chen et al., 1985). Based on the qualitative nature of the effects seen, the relatively long retention of PCDFs compared to that of PCBs in the tissues of the victims, and the comparative toxicity of samples of the contaminated oil and pure PCDFs in rats and monkeys, it is now believed that PCBs alone could not have been responsible for the signs and symptoms exhibited by the Yusho and Yu-Cheng victims (Kashimoto et al., 1981; Kunita et al., 1985; Kashimoto and Miyata, 1987). Rather, the effects seen were most likely due to the presence of PCDFs, specifically penta- and hexachlorinated congeners, in the contaminated oil. Thus, studies among the populations exposed to contaminated rice oil in Japan and Taiwan are useful in providing information on the human health effects of these PCDFs and closely related halogenated aromatic hydrocarbons.

### *2.3 Conclusions on Definite or Probable Effects*

The only effect that can definitely be linked to PCDD exposure is chloracne. The evidence from the Yusho and Yu-Cheng episodes indicates that chloracne and the ocular effects were probably caused by PCDFs. Evidence that other findings among the victims were due to PCDFs is relatively weak.

## 3 Effects Possibly Caused by PCDDs/PCDFs

### *3.1 Porphyria*

Porphyria, a disorder of heme synthesis, has been found in experimental animals treated with 2,3,7,8-TCDD (Kociba et al., 1976, 1978; Sweeney et al., 1984). The biochemical lesion involved is presumed to be inhibition of uroporphyrinogen decarboxylase. It is manifested by an accumulation of porphyrins in the liver and altered patterns of urinary porphyrin excretion (Sweeney et al., 1984). A specific form of human porphyria, known as porphyria cutanea tarda (PCT), in which discoloration and increased fragility of the skin and sometimes hirsutism accompanies the altered urinary porphyrin excretion pattern, may also occur.

Abnormal urinary porphyrin patterns without the accompanying skin lesions of PCT have been observed in individuals exposed to mixtures containing 2,3,7,8-TCDD. Doss and colleagues (1984) reported that 13 of 60 people studied from the area contaminated by 2,3,7,8-TCDD in Seveso, Italy had abnormal urinary porphyrin excretion patterns one year after the accident, and 10 of these had returned to normal three years later. Studies of several other populations that may have been exposed to 2,3,7,8-TCDD, including the Ranch Hand cohort (Lathrop et al., 1984) and a cohort from Missouri (Webb et al., 1984), revealed no abnormalities in urinary porphyrin excretion patterns.

Porphyria cutanea tarda has been reported among workers who were accidentally exposed to complex chemical mixtures containing 2,3,7,8-TCDD. Jirasek and coworkers (1973, 1974) reported PCT in 11 of 78 workers from a plant in Czechoslovakia where the workers were engaged in the manufacture of phenoxy herbicides and hexachlorobenzene. In another plant in New Jersey, 11 of 29 workers examined had PCT (Bleiberg et al., 1964). PCT was also observed in workers involved in the manufacture of phenoxy herbicides at a plant in Nitro, West Virginia (Suskind, 1983). In each of these populations, reexamination of workers at a later date indicated that the skin lesions associated with PCT had disappeared and only a few abnormalities in urinary porphyrin excretion patterns remained (Poland et al., 1971; Pazderova-Vejlupkova et al., 1981; Suskind and Hertzberg, 1984). In a review of these and other reports, Hobson (1984) concluded that the PCT seen in these populations was most likely attributable to exposure to hexachlorobenzene. A study of a family exposed to 2,3,7,8-TCDD following the Seveso accident led to the suggestion that there was a genetic predisposition to PCT which was aggravated by exposure to alcohol or dioxin (Strik et al., 1980; Doss et al., 1984). PCT has not been a finding in studies of other human populations that were exposed to 2,3,7,8-TCDD in the absence of hexachlorobenzene and trichlorophenol.

### *3.2 Changes in Blood Lipids and Indicators of Liver Function*

Various studies have reported changes in blood lipids and liver function tests in TCDD-exposed workers. A study by Martin (1984), 10 years after an industrial accident in the United Kingdom, reported increases in serum cholesterol and triglyceride concentrations in workers both with (Donovan et al., 1984) and without (Axelson, 1980) chloracne. The average cholesterol and triglyceride levels were approximately 10 and 37% higher, respectively, than those for controls. Differences were also found for D-glucaric acid/creatinine ratios in urine samples, with exposed workers having levels about 30% higher than among controls. Again, there were no differences for workers with and without chloracne. Measurements were also taken of serum gamma-glutamyl transferase (GGT) activity, but while they correlated with triglyceride levels, the differences between exposed and unexposed workers were very small.

These results are puzzling because there were no differences found between workers with and without chloracne, and presumably those with chloracne would have experienced higher exposures, on average, than those without chloracne. However, a study on a similar group of 206 workers exposed at a U.S. trichlorophenol plant found triglyceride levels of those with chloracne to be 16% higher than workers without chloracne ($p = 0.056$). The GGT levels were 50% higher ($p = 0.003$), and serum glutamic-oxaloacetic transaminase (SGOT) levels were 15% higher ($p = 0.002$) in those with chloracne than in those without chloracne (Moses et al., 1984). No differences were found in serum cholesterol levels. A separate study of a larger number of workers from the same plant found no differences between exposed and non-exposed workers for either total cholesterol or triglycerides (Suskind and Hertzberg, 1984). In addition, another study among trichlorophenol plant workers and a 2,4,5-T manufacturing cohort failed to find differences in acitivity of liver enzymes, including lactate dehydrogenase, aspartate transaminase, alanine transaminase, and alkaline phosphatase (Bond et al., 1982).

A study of children exposed to 2,3,7,8-TCDD at Seveso, Italy, reported that boys with the highest exposures had increased GGT and alanine aminotransferase activity, but no significant differences were observed in girls (Mocarelli et al., 1986). The differences among boys were described as small and disappeared with time. Additional studies demonstrated elevated D-glucaric acid levels in urine collected from Seveso children (Ideo, 1984). Children with chloracne had levels nearly twice as high as children without chloracne ($p < 0.05$). Furthermore, adults ($n = 117$) in Seveso had levels 37% higher than adults ($n = 117$) from a control area ($p < 0.05$). On the other hand, the Missouri study reported no differences in urinary D-glucaric acid excretion between the exposed and comparison subjects (Hoffman et al., 1986).

### *3.3 Gastrointestinal Ulcer*

A study of trichlorophenol plant workers reported that those exposed to 2,3,7,8-TCDD were four times more likely to report a history of GI tract ulcer than those not exposed (Suskind and Hertzberg, 1984). After removal of subjects with symptoms and with known causes for their ulcers before exposure, there remained a difference between the groups with 16.6% (30 cases) reporting GI tract ulcers among the exposed and 5.5% (9 cases) among the non-exposed ($p < 0.01$). Ten exposed subjects reported surgical treatment for their ulcers, compared to none among the non-exposed.

Similar findings were reported in a study at another trichlorophenol manufacturing plant (Bond et al., 1982). There was a significantly greater prevalence of diseases of the digestive system in the 2,4,5-T cohort compared with a control cohort (17 out of 135 men examined in the 2,4,5-T cohort, compared with 27 out of the 540 in the control cohort, prevalence ratio = 2.5, 90% confidence interval 1.5–4.2).[2] Some of the difference was due to X-ray diagnosed ulcers with 9 cases among 87 in the 2,4,5-T cohort (prevalence ratio 2.1, 90% confidence interval 1.1–4.2). However, the 27 workers involved in a chloracne incident in an area of trichlorophenol production, and who probably experienced higher exposure to PCDDs than the 2,4 5-T cohort, did not manifest a higher prevalence of ulcers compared to their control group. (Two cases were observed; approximately 1.75 were expected on the basis of their control cohort; crude estimate of prevalence ratio = 1.14.) While this weakens the evidence for a link with PCDDs, the numbers in the trichlorophenol cohort were too small to attach much importance to the absence of an association.

These two independent studies provide some evidence that exposure to 2,3,7,8-TCDD may cause upper GI tract ulcers, in spite of the lack of effects in a small high-exposure subgroup. Coupled with the hyperplasia observed in the stomach of exposed monkeys (see Risk Characterization Working Group, this volume), the evidence suggests that exposure to 2,3,7,8-TCDD may contribute to the development of GI tract ulcers in humans.

### *3.4 Immune System Effects*

There have been very few reports of immune system toxicity from human exposure to PCDDs (Knutsen et al., 1987). However, there are some immune system studies of residents in the Quail Run Mobile Home Park who were potentially exposed to 2,3,7,8-TCDD due to use of contaminated waste sludge to suppress dust on a dirt road (Knutsen, 1984; Hoffman et al., 1986; Stehr et al., 1986). While there were problems in assuring accuracy of observations, the exposed group was reported to have had a significantly increased frequency of anergy compared to controls (11.8% vs 1.1%), and of relative anergy (35.3% vs 11.8%). There were also small differences in abnormal T-cell subset test results (10.4% vs 6.8%), T4/T8 ratios of less than 1 (8.1% vs 6.4%), and an abnormality in the functional T-cell test results (12.6% vs 8.5%), but none of the T-cell findings were statistically significant. The authors interpreted the findings as evidence that 2,3,7,8-TCDD exposure was associated with depressed cell-mediated immunity.

A follow-up study was subsequently conducted involving participants who were initially anergic or relatively anergic, with 43 (56%) participating. The

[2] Confidence intervals are reported for estimates quoted in this review; if authors did not give confidence intervals, these were calculated, where possible, from the published data

findings are available from a meeting abstract (Evans et al., 1987). The repeat skin tests indicated that none of the participants were anergic, and only one exposed subject and one unexposed subject were relatively anergic. The only T-cell measures outside the normal range were the percentage of T4 cells and T4/T8 cell ratios in the exposed group. the authors concluded that, among several possible interpretations for the changes in skin test results, recovery from the effects of 2,3,7,8-TCDD was the least plausible. In view of the weakness of the initial T-cell findings, and the lack of evidence regarding anergy, an effect on the immune system is questionable.

### *3.5 Neurotoxicity*

In a study of 470 people exposed at Seveso, it was noted that among 55 subjects with chloracne or abnormal serum hepatic enzyme levels, the prevalence of peripheral neuropathy was 2.8 times (95% confidence interval 1.2–6.5) that for persons without such manifestations of exposure (Filippini et al., 1981). The changes in nerve conduction velocity applied to a small number of 2,3,7,8-TCDD exposed subjects only and it was not possible to rule out other causes (Reggiani, 1983).

Neurotoxic effects involving polyneuropathy of the lower extremities were reported following industrial exposure in Czechoslovakia (Pazderova-Vejlupkova et al., 1981). Among the 55 exposed individuals, 52 of whom developed chloracne, 23% were diagnosed as having polyneuropathy when first exposed. A later examination of 44 subjects found 31% with polyneuropathy.

However, these findings have not been replicated in studies of other workers exposed after industrial accidents (Suskind, 1985). One study reported 11 out of 60 chloracne cases had decreased sensation to pin prick, compared to none of 34 subjects who never had chloracne ($p < 0.01$), but there were no other significant neurological differences between the two groups (Moses et al., 1984).

No neurological differences were found between the exposed and non-exposed subjects in the Missouri study described in the section above (Hoffman et al., 1986).

### *3.6 Conclusions on Possible Effects*

Although there have been independent reports of PCT and/or alterations in porphyrin metabolism among humans exposed to complex chemical mixtures containing 2,3,7,8-TCDD, these effects appear to be reversible and are associated only with exposures that result in another, more common manifestation of toxicity, i.e., chloracne. Furthermore, the systemic as well as cutaneous manifestations of porphyria may be attributable to other components of the mixtures.

It would also appear that this class of effects is the result of short-term exposures to relatively high concentrations of the causative agent.

Findings concerning individual effects measured by liver function tests, blood lipid levels, urinary porphyrin, and D-glucaric acid excretion are inconsistent from study to study and have therefore been classified as possible rather than probable effects of PCDDs. However, it is noteworthy that studies involving heavy exposures, such as with industrial accidents and at Seveso, have usually reported at least some effects which indicate that the liver may be a target organ in humans. Thus, PCDD exposure, in particular 2,3,7,8-TCDD, may cause various changes involving the liver although there is no evidence that these changes are associated with adverse health effects.

On the basis of two independent studies, it is possible that 2,3,7,8-TCDD can contribute to the development of upper gastrointestinal tract ulcers. The evidence that PCDDs have caused immune system effects in humans is weak, and there is no good evidence that PCDD exposure can cause anergy. The findings in one study of effects on human T-cells are consistent with animal studies. However, even if these effects are real, it is not known if they would have any health implications. Finally, the overall evidence does not provide convincing support to the hypothesis that PCDDs have caused neurologic effects in humans.

# 4 Effects on Reproduction

Reproductive outcomes which warrant consideration include infertility, miscarriage and spontaneous abortion, and congenital anomalies and developmental abnormalities. The human studies which are relevant to potential reproductive effects of phenoxyacids include essentially ecological studies with no evidence of individual maternal or paternal exposure, and occupational studies with individual information on exposures which may have occurred in a manufacturing setting or in the use of herbicides. Their findings generally indicate no reproductive effects of exposures at the levels involved.

## *4.1 Studies With Ecological Exposure Data*

### General Populations

Field and Kerr (1979) identified the annual usage of 2,4,5-T in Australia between 1965 and 1976 and assessed its relationship to the following year's prevalence rate of neural tube defects at birth in New South Wales. They found a positive correlation between the two, with neural tube defects increasing from 1.72 to

2.30 per 1000 births, while 2,4,5-T use increased from 90 to 482 metric tons per year. Such a correlation was not found in Hungary for neural tube defects, which declined a little during a period when 2,4,5-T use increased from 28 to 660 metric tons per year, nor was there a correlation with rates of stillbirths, cleft lip, cleft palate, and cystic kidney disease (Thomas, 1980).

A study in Arkansas of 1201 cases of cleft lip and/or cleft palate involved dividing the state into high, medium, and low use of 2,4,5-T between 1948 and 1974 on the basis of rice acreage (Nelson et al., 1979). No significant differences in facial cleft defect rates were found between these areas.

The U.S. Environmental Protection Agency (1979) investigated spontaneous abortion rates in three Oregon areas in relation to 2,4,5-T spray practices. Significantly higher rates of spontaneous abortion were found in the area in which 2,4,5-T was used, but the findings have been questioned because of inadequate methods for case ascertainment and other problems with the study (Coulston and Olajos, 1980).

The possible effects of aerial spraying were studied in the Northland regions of New Zealand by dividing it into seven areas according to the extent of 2,4,5-T spraying as assessed by a detailed review of the records of the companies involved (Hanify et al., 1981a,b). An incidence ratio of 1.73 (90% confidence interval 1.44–2.08) was calculated comparing birth malformation rates after aerial spraying of 2,4,5-T commenced with rates before aerial spraying began. In addition, an association with the extent of 2,4,5-T use was found for all birth malformations combined and for talipes, hypospadias and epispadias, and heart defects separately. No significant associations were found for neural tube defects, nor for cleft lip and palate. A significant negative association was found with stillbirths (Hanify et al., 1981a,b). It should be noted that multiple comparisons were involved and that the study was basically ecological in nature since maternal exposure was determined only by geographic area of residence.

## Veterans of Vietnam Conflict

Extensive anecdotal information centers around the issue of whether or not veterans with potential exposure to Agent Orange, a mixture of 2,4,5-T and 2,4-D contaminated with 2,3,7,8-TCDD, have experienced a variety of effects, including reproductive effects such as birth anomalies and spontaneous abortion. A 1984 review concluded that there was no scientific evidence which indicates that men who were previously exposed to Agent Orange were at increased risk of having children with birth defects, but noted that available data were inadequate to assess this possibility critically (Friedman, 1984).

Two important case-control studies have since been published. An Australian case-control study involving 8517 case-control pairs found a relative risk of 1.02 (95% confidence interval 0.78–1.32) for veterans fathering children with birth anomalies compared to non-veterans (Donovan et al., 1984). A similar U.S. study found an overall relative risk of 0.97 (95% confidence interval 0.83–1.14)

(Erickson et al., 1984). As might be expected, some positive associations were found for specific birth defects, and some negative associations. For example, in the U.S. study, increasing risks for spina bifida were estimated for increasing values of the exposure index for Agent Orange. However, the trend for anencephalus was in the opposite direction. Similarly, a positive trend for cleft lip without cleft palate was counterbalanced by a negative trend for cleft palate. While these studies cannot dispel the possibility that rare birth anomalies could be caused by male exposure to Agent Orange, they make it seem unlikely that Agent Orange caused birth anomalies among the offspring of exposed males (Dan, 1984).

Studies among Vietnamese themselves are difficult to interpret in the light of problems of case ascertainment and exposure assessment, and the many potential confounding factors associated with a country at war. Studies have suggested a link between Agent Orange exposure and a variety of reproductive effects including infertility, spontaneous abortions, birth defects, and hydatidiform mole (Hatch, 1984; Constable and Hatch, 1985; Sterling and Arundel, 1986a). However, the lack of systematically organized analytical epidemiology studies limits these findings to a classification of hypothesis-generating observations.

## Seveso, Italy

An industrial accident in July of 1976 released toxic quantities of 2,3,7,8-TCDD resulting in exposure to the surrounding community. Limited information is available concerning reproductive effects. One study compared congenital anomaly rates in the Seveso region with those in the unexposed Lombardy region (Bruzzi, 1983). Results were mixed, with differences in both directions. Rates of neural tube defects were higher in the Seveso region than in the Lombardy region (17.9 per 10,000 births compared with 10.4 per 10,000 births, $p = 0.06$). The same was true for hypospadias (28.2 per 10,000 births in the Seveso region vs 2.3 per 10,000 births, $p < 0.001$). However, looking within the Seveso region alone, the variation in these rates with the extent of potential exposure, judged by area of residence, was weak. In particular, only 3 anomalies of the genital organs (presumably hypospadias) were found in the high exposure region, whereas there were 27 such anomalies in the area outside the exposure zones. Thus, the hypospadias rate in the non-exposed area around Seveso was also much higher than in the comparison Lombardy region. No positive evidence regarding congenital anomalies can therefore be drawn from this study.

The most recent report concerning birth defects related to the Seveso incident also failed to demonstrate increased risks of birth defects associated with 2,3,7,8-TCDD exposure (Mastroiacova et al., 1988). The relative risk estimate for major defects among 461 births to mothers resident in the contaminated zones (A and B) compared to 12,391 births to mothers living outside the area was 1.02 (90% confidence interval 0.64–1.61). No major defects were reported for the 26 births to mothers resident in the most contaminated zone

(zone A). Only 1 of the 435 babies born to mothers in the second most heavily contaminated zone (zone B) had a neural tube defect, and none had facial clefts. However, 3 babies had congenital heart defects (1.7 expected based on births outside the area), and 4 had hypospadias (1.4 expected), one more than found in the earlier report reviewed above. In light of the overall lack of increase in major defect rates, the findings concerning heart defects and hypospadias could well be the result of more intensive examination of those exposed which could lead to an observational bias.

Another study following the Seveso incident involved tests of maternal blood, and specimens from the placenta, umbilical cord, and aborted fetuses from the Seveso area of 16 exposed women, some of whom elected to have abortions (Tenchini et al., 1983). A problem with this study was that control group specimens did not become available until over 2 years later when abortion in Italy was legalized. The authors reported statistical analyses which did not, as a whole, show any effect of 2,3,7,8-TCDD exposure. One exception was that fetal tissue from exposed pregnancies showed significantly higher aberrant cell frequencies compared with controls, but it was suggested that the most plausible explanation was factors inherent in growth in culture.

### *4.2 Studies Identifying Exposed Individuals: Herbicide Spraying*

A pregnancy outcome study of wives of professional herbicide sprayers has been conducted in New Zealand (Smith et al., 1981). The phenoxy herbicides 2,4,5-T and 2,4-D were the main herbicides used by these sprayers. There were 1172 births among applicator families in the study period, and 1122 births among a comparison group of agricultural contractors. Information was gained by postal questionnaire with an overall response rate of 86%. Major congenital defects were reported for 2% of applicator births, a figure similar to that for other New Zealand studies, and 1.6% of agricultural contractor births. The difference was not significant. The stillbirth rates and miscarriage rates were virtually the same. A further analysis selected those pregnancy outcomes associated with spraying 2,4,5-T by the father in the same year or the year before and compared them with pregnancy outcomes not associated with spraying in this period of time (Smith et al., 1982a). The relative risk for congenital defects among the exposed was 1.19 (90% confidence interval 0.58–2.45) and the miscarriage relative risk was 0.89 (90% confidence interval 0.61–1.30).

### *4.3 Studies Identifying Exposed Individuals: Chemical Manufacture*

A study using interviewer-administered questionnaires was conducted among 715 wives of employees of a chemical company in Michigan with a comparison of birth outcomes for wives of employees with potential for exposure to dioxin and birth outcomes for wives of employees with no exposure to 2,3,7,8-TCDD (Townsend et al., 1982). Exposures were sufficient for some workers to have had chloracne. The outcomes examined included spontaneous abortions, stillbirths, infant deaths, and congenital malformations with no significant associations being found. The odds ratio estimate for congenital malformations was 1.08 (95% confidence interval 0.63–1.83).

A morbidity study of workers manufacturing 2,4,5-T in Nitro, West Virginia, also included information concerning birth outcomes (Suskind and Hertzberg, 1984). The study population included 204 clearly exposed workers (56% still had clinical evidence of chloracne), and 163 not exposed. Reproductive information was obtained from the male employees and included 655 pregnancies classified as exposed, and 429 classified as not exposed. Both stillbirths and deaths within 4 weeks occurred more frequently among the exposed but not significantly so [28 stillbirths plus deaths within 4 weeks associated with 655 pregnancies among the exposed compared with 11 out of 429 pregnancies among the non-exposed (relative risk = 1.7, $p = 0.19$)]. The miscarriages rate was a little lower among the exposed, and the birth defect prevalence was about the same in the two groups.

### *4.4 Conclusions Concerning Reproductive Effects*

While there are suggestions of individual findings, there are no consistent associations in a variety of studies of pregnancy outcomes among wives of exposed males. These findings suggest no reproductive effects at the levels of human exposure involved. There are no studies involving female exposure in the work setting, and studies involving environmental exposure of women are, by and large, uninformative.

## 5 Carcinogenicity Studies in Humans

### *5.1 Introduction*

Human studies relevant to the carcinogenicity of dioxins and furans involve a variety of cancer sites. Particular attention has been given to soft tissue sarcoma and malignant lymphoma since there are several studies relating them to exposure to phenoxy herbicides. The search for soft tissue sarcomas and lymphomas has extended to chemical industry workers, particularly workers who may have been exposed to 2,3,7,8-TCDD, since it has been shown to be an animal carcinogen. A variety of studies address tumors of other sites. This section, therefore, deals with three main topics: soft tissue sarcoma, malignant lymphoma, and tumors of other sites.

## 5.2 *Soft Tissue Sarcoma*

### Exposure to Phenoxyacetic Acid Herbicide Spraying

*Studies in Sweden.* The first soft tissue sarcoma case-control study followed the observation of a number of patients at a cancer clinic in Sweden who reported previous exposure to phenoxyacetic acid herbicides (Hardell and Sandstrom, 1979). A total of 52 male patients, 21 living and 31 deceased, were identified from records of the Department of Oncology of the University Hospital of Umea as having been admitted with a diagnosis of soft tissue sarcoma between 1970 and 1977. Four matched controls for each case were selected from the National Population Registry for living patients or from the National Registry for Causes of Death for deceased patients. Ascertainment of exposure was by telephone and mailed questionnaire. A relative risk estimate of 5.7 was calculated for exposure of more than 1 day to phenoxyacetic acids and/or chlorophenols (95% confidence interval 2.9–11.3).

When patients and controls with exposure to chlorophenols were excluded, the relative risk estimate was 5.3 (95% confidence interval 2.4–11.5) with 13 patients exposed to various phenoxy acids. Of the 13, 12 had reported exposure to 2,4,5-T or 2,4-D, and one to 4-chloro-2-methylphenoxyacetic acid (MCPA) alone. Combined exposure to 2,4,5-T and 2,4-D was reported in 9 cases. The exposures of 2 patients solely consisted of working on ground which was wet from earlier spraying. Latency from first exposure was predominantly in the range 10–20 years. The duration of exposures was short; the median duration of exposure was between 3 and 4 months, with only 2 individuals reporting more than 12 months of exposure.

A second study of soft tissue sarcoma was then undertaken, this time in southern Sweden, to seek confirmation of the earlier findings. The study involved 110 cases, and 2 controls were selected per case by the same methods used in the first study (Eriksson et al., 1981). The relative risk estimate for exposure to phenoxy acids or chlorophenols was 5.1 (95% confidence interval 2.5–10.4). A relative risk estimate of 8.5 was obtained for exposure to phenoxy acids of greater than 30 days and 5.7 for exposures of less than or equal to 30 days.

A possible explanation of the findings involved the contamination of 2,4,5-T with 2,3,7,8-TCDD, a potent carcinogen in rats and mice. However, a relative risk estimate of 4.2 was calculated for exposure to phenoxy herbicides other than 2,4,5-T, casting some doubt on this hypothesis.

It was also possible that greater recall of past herbicide exposure by soft tissue sarcoma patients than by controls may have produced the earlier findings. Because of this possibility, a study was conducted involving colon cancer patients (Hardell, 1981a). The study involved 157 male colon cancer patients who were interviewed in the manner of the earlier studies and whose exposure was compared with that of the controls in the earlier studies. A relative risk estimate of 1.3 (95% confidence interval 0.6–2.8) was obtained. The absence of an association with colon cancer reduces the likelihood of significant recall bias in the earlier studies; although, the assumption of comparable recall in the two studies would only be valid if colon cancer patients were age matched with soft tissue sarcoma patients. If not, colon cancer patients, who would more likely be older, may have diminished recall of long past exposure.

A reanalysis of the earlier data using a method proposed by Axelson (1980) did not suggest recall bias had occurred. The basis of this method is that recall bias should vary by occupation. If there is over-recall of exposure in one particular occupational group, then the risk ratio for those not reporting exposure should be reduced when compared with those not reporting exposure in other occupational groups. This was not the case for the agricultural and forestry occupational group in the soft tissue sarcoma studies. The risk ratio for those not reporting exposure to phenoxy herbicides or chlorophenols was 1.2 compared to the non-exposed in other occupations. Since the risk ratio was not reduced below 1, it was again thought that recall bias had not occurred. However, it should be noted that this test is based on several assumptions, one being that there is no recall bias in terms of occupation in agriculture and forestry.

More recent studies in Sweden have involved linking occupational data from the census with mortality data (Wiklund, 1986; Wiklund and Holm, 1986). Risks of soft tissue sarcoma death were examined for the 354,620 Swedish men who were employed in agriculture or forestry according to the 1960 census. Only 13% of these men were not working in agriculture or forestry at the time of the 1950 census. The reference cohort was 1,725,845 men employed in other industries. A total of 331 cases were observed in the study cohort yielding a relative risk estimate of 0.9 (95% confidence interval 0.8–1.0). Exposure to 2,4,5-T and 2,4-D was thought to be mainly in silviculture. The relative risk estimate for the silviculture sub-cohort was 0.8 (95% confidence interval 0.3–19).

While individual exposure data are lacking from the Swedish cohort studies using census data, the results are nevertheless inconsistent with the case-control studies. Wiklund and Holm (1986) estimated that 15% of agricultural and forestry workers had been exposed to phenoxy acids. They calculated an expected relative risk of 1.5 for soft tissue sarcoma death in these occupations based on Hardell's (1981a) reported odds ratio of approximately 6.0 for persons exposed to phenoxy acids for less than 1 year. It is remarkable that the census linkage study failed to report any increased risk for all agriculture and forestry workers, and that the upper 95% confidence limit for the relative risk estimate was 1.0. These findings raise questions concerning the validity of Swedish case-control study results, which, as far as we are aware, have not been satisfactorily resolved.

*Studies in New Zealand.* The initial Swedish studies have stimulated several further studies of soft tissue sarcoma and phenoxy herbicide spraying. An initial analysis of occupations recorded on the National New Zealand Cancer Registry did not find an excess of soft tissue sarcoma cases in agriculture and forestry compared with patients with cancer of other sites on the cancer registry (Smith et al., 1982b). Telephone interviews of past occupations and specific use of phenoxy herbicides were completed for 82 subjects with soft tissue sarcoma who were recorded on the Cancer Registry between 1976 and 1980 and 92 controls with cancers of other sites (Smith et al., 1983, 1984). A relative risk estimate of 1.6 (90% confidence interval 0.7–3.3) was calculated for those who were probably or definitely exposed to phenoxy herbicides for more than 1 day but not in the 5 years prior to cancer registration. However, the relative risk decreased to 1.3 (90% confidence interval 0.6–2.9) for probable or definite exposure for 5 days or more but not in the 10 years prior to cancer registration.

A second study was undertaken with interviews of 51 further cases appearing more recently on the Cancer Registry up to 1982 (Smith and Pearce, 1986). An estimate of relative risk for exposure of more than 1 day not in the 5 years prior to registration with cancer of 0.7 (90% confidence interval 0.3–1.5) was obtained. The combined estimate for 133 cases registered between 1976 and 1982 was 1.1 (90% confidence interval 0.7–1.8).

The herbicide 2,4,5-T has been used extensively in New Zealand since 1948 (Smith et al., 1983). Other phenoxy herbicides have been used such as 2,4-D, which was often used in combination with 2,4,5-T. The phenoxy acids, MCPA and methylchlorophenoxybutyric (MCPB), have also been used. Ground spraying has mainly been by professional applicators. Since these professional applicators spray phenoxy herbicides for 6 or more months per year, and since some have sprayed for 20 years or more, the fact that none were found with soft tissue sarcoma in the case-control studies is worthy of note (Smith, 1983). Ground spraying commenced in the late 1940s. Cases were ascertained between 1976 and 1982, so there is ample latency. The level of contamination with 2,3,7,8-tetrachlorodibenzo-*p*-dioxin has been greatly reduced in recent years, but was of the order of 1 ppm in 1971, with levels prior to that presumably the same or higher (Smith and Pearce, 1986). The level of contamination of 2,4,5-T used in Sweden was probably similar, since measurements of old samples of 2,4,5-T revealed levels of 2,3,7,8-TCDD on the order of 1 ppm (Norstrom et al., 1979).

The New Zealand studies involved use of other cancer patients as controls. However, a comparison of such controls with general population controls also interviewed by telephone did not reveal differences in past exposure prevalence (Pearce et al., 1986). Therefore, the use of other cancers as controls does not explain the difference in findings between the Swedish and New Zealand studies.

*United States Case-Control Studies.* Two case-control studies conducted in the United States have not found an association between phenoxy herbicide use and soft tissue sarcoma. A telephone interview study in Kansas included 95 patients with soft tissue sarcoma who were compared with randomly selected controls from the general population (Hoar et al., 1986). Farm herbicide use was reported by 22 patients, yielding an odds ratio of 0.9 (95% confidence interval 0.5–1.6).

The phenoxy herbicide 2,4-D was used extensively in Kansas, so this study contributes to the evidence against the soft tissue sarcoma phenoxy herbicide hypothesis (Colton, 1986).

In a second case-control study of 128 soft tissue sarcomas from the State of Washington, the estimated relative risk for occupational exposure to phenoxy herbicides was 0.8 (95% confidence interval 0.5–1.2). Interestingly, study subjects with Scandinavian surnames were identified, and the relative risk estimate for phenoxy herbicide exposure from the 15 soft tissue sarcoma patients with Scandinavian surnames was 2.8 (95% confidence interval 0.5–15.6). However, taken overall, the study provides evidence against an association between phenoxy herbicide exposure and soft tissue sarcoma (Woods et al., 1987).

*An Italian Case-Control Study.* A case-control study of soft tissue sarcoma and exposure to phenoxy herbicides focused on exposure of rice weeders, who may have had skin contact with phenoxy herbicides during weeding activities (Vineis et al., 1986). The study involved both men and women including 68 with soft tissue sarcoma and 158 referents. No men with soft tissue sarcoma had been exposed. Among the living subjects, the odds ratio for males was 0.91 (90% confidence interval 0.21–3.91). For females, the odds ratio was 2.42 (95% confidence interval 0.59–12.37), based on 5 exposed soft tissue sarcoma cases. Three of these women, as compared to one referent, were exposed during 1950–1955, a period during which exposures may have been heavy according to the authors. It should be noted that rice weeders were considered to have been "exposed" in this study when they worked weeding rice after 1950 and did not work exclusively in a small rice allotment of their own. It did not necessarily mean they had any direct contact with phenoxy herbicides.

*Vietnam War Veteran Studies.* In general, studies of soft tissue sarcoma in relation to the Vietnam War use of Agent Orange, a mixture of 2,4,5-T and 2,4-D, are inconclusive. One study ascertained Vietnam service for 281 men with soft tissue sarcoma and a matched control group (Greenwald et al., 1984). Cases were diagnosed between 1962 and 1980 and identified on the New York State Cancer Registry. The relative risk estimate for service in Vietnam was 0.53 (95% confidence interval 0.21–1.31), with 10 cases reporting Vietnam service. A relative risk estimate of 0.70 was obtained for reported contact with Agent Orange, dioxin, or 2,4,5-T. However, since the Vietnam War started in 1962 and finished in 1971, and since diagnosis was by 1980, the elapsed time may not be sufficient to make any firm conclusions about phenoxy acid herbicides and soft tissue sarcoma.

A case-control study was undertaken of 234 Vietnam-era veterans who served in the U.S. military between 1964 and 1975, and who were treated at a Veterans Administration Hospital between 1969 and 1983 with a diagnosis of soft tissue sarcoma (Kang et al., 1986). The control group consisted of 13,496 patients sampled from the same Vietnam-era patient population from which the cases were drawn. The odds ratio estimate for previous service in Vietnam was 0.83 (95% confidence interval 0.63–1.09). Furthermore, there was no trend with year of hospital discharge; those with hospital discharge dates between 1976 and 1980, and between 1981 and 1983, showed no increase in the percentage with Vietnam service.

A recent case-control study involved 217 cases selected from the Armed Forces Institute of Pathology which were compared with 599 controls from the patient logs of referring pathologists or their pathology departments (Kang et al., 1987). The odds ratio estimate for service in Vietnam was 0.85 (95% confidence interval 0.54–1.36). The odds ratio was 2.57 (95% confidence interval 0.72–9.36) for those who had a combat related military occupation specialty code, based on 8 exposed cases. The odds ratio estimate further increased when their location was within Military Region III, the area where Agent Orange spraying was reported to be extensive, with an estimate of 8.64 (95% confidence interval 0.77–111.84), based on 5 exposed cases. The authors noted the lack of study power when considering these sub-groups and concluded that the possibility of a modestly increased risk of soft tissue sarcoma associated with Agent Orange exposure in Vietnam among selected groups of Vietnam veterans could neither be confirmed nor ruled out by their study. It should also be noted that over 80% of the cases were observed less than 10 years after the last troops were exposed to Agent Orange in Vietnam.

There have been several state health department studies concerning veterans of the Vietnam conflict and mortality from soft tissue sarcoma. A study of these veterans in the State of New York reported a relative risk estimate of 1.15 (95% confidence interval 0.61–2.17) for soft tissue sarcoma deaths in veterans compared with non-veterans (Lawrence et al., 1985). By contrast, a study for the State of Massachusetts reported a relative risk estimate of 4.73 (95% confidence interval 2.16–8.99) based on 9 veteran cases (Kogan and Clapp, 1985). In addition, a study in West Virginia reported a relative risk estimate of 4.3 (95% confidence interval 0.9–12.7), but was based on only 3 cases (West Virginia, 1986). According to a follow-up report made to the Veterans Administration Committee on Environmental Hazards, two of the three with soft tissue sarcomas were in veterans who never served in Vietnam; one served in Thailand and one served on board a ship off the coast of Vietnam (Minutes of a Meeting of the Committee on Environmental Hazards, Veterans Administration, Washington, D.C., March 4, 1986). It should also be noted that these two studies reporting an apparent association with mortality from soft tissue sarcoma were much smaller than the studies reported above which did not find such an association.

The highest exposures among Vietnam-era veterans were experienced by the Ranch Hand personnel who sprayed Agent Orange (Gross et al., 1984). Follow-up examinations of the Ranch Hand Group (n = 1016) and a matched group of Vietnam War veterans who were not in Ranch Hand (n = 1293) were completed between January 1985 and September 1987 (Lathrop et al., 1987). One case of soft tissue sarcoma was reported in each group. Thus, there is no evidence at this stage of follow-up for any increased risk of soft tissue sarcoma for the most heavily exposed veterans.

## Other Occupational Exposures

*Soft Tissue Sarcoma in the Chemical Manufacturing Industry.* The first stage of the chemical manufacture of 2,4,5-T involves the production of trichlorophenol. Workers in trichlorophenol plants have potential exposure to a variety of chemicals including 2,3,7,8-TCDD. Since exposure to 2,3,7,8-TCDD had in the past been sufficient to cause chloracne among some workers, the finding of soft tissue sarcoma in phenoxy herbicide sprayers suggested searching for cases among trichlorophenol plant workers, even though the expected number of cases would be very small. Reports of seven cases among small cohorts of workers gave the impression that there was an excess of soft tissue sarcomas among such workers (Honchar and Halperin, 1981). Only two of the seven cases have subsequently been confirmed as soft tissue sarcomas, and as having workplace exposure in the trichlorophenol plant areas (Fingerhut et al., 1984). The evidence from these studies is therefore weak and inconclusive.

Recently, information has come from a large chemical industry cohort study involving 2192 workers who had potential occupational exposures to chlorinated dioxins (Ott et al., 1987). There was one reported death from soft tissue sarcoma with 0.4 expected, but the case was one which was reclassified to renal clear-cell carcinoma after pathological review.

However, an excess of soft tissue sarcomas has been discovered in a cohort study of workers involved in the manufacture of phenoxy herbicides in Denmark (Lynge, 1985). This study involved linkage of worker identification with data in the Danish Cancer Registry. Among males, 5 cases of soft tissue sarcoma were found when 1.8 would have been expected (RR = 2.72, 95% confidence interval 0.88–6.34). No cases were discovered among females (0.75 expected). For more than 10 years latency since first exposure, the relative risk estimate for males was 3.67 (95% confidence interval 1.00–9.39) involving 4 cases. This study is of some importance in the context of PCDD exposure and soft tissue sarcoma since 4 of the 5 soft tissue sarcoma cases worked in areas where they were relatively unlikely to be exposed to PCDDs. In fact, 3 worked in the shipping department where they mainly handled sealed goods. The evidence from this cohort does not, therefore, implicate PCDDs with soft tissue sarcoma.

## Community Exposure: Seveso, Italy

A review of soft tissue sarcoma cases found between 1975 and 1981 yielded 15 among residents in the "polluted area" with an age adjusted incidence rate of 5.73 per 100,000 inhabitants (Puntoni et al., 1986). There were 44 cases in the unpolluted area yielding an incidence rate of 3.34 per 100,000. However, the latency period was very short; the authors later noted that latency considerations as well as exposure definition deficiencies do not permit an assessment of causality (Merlo and Puntoni, 1986). One possibility is that community exposures may have preceded the industrial accident (Hardell and Eriksson, 1986; Puntoni et al., 1986). It is noteworthy that there was no increase in soft tissue sarcoma incidence rates between 1975 and 1981; in fact, the highest rate, which was 8.47 per 100,000 was for 1975, the year before the accident. It therefore seems most unlikely that this apparent elevation in incidence in the polluted area could relate to the accident. Exposures before the accident are unlikely to have been high enough to have been of concern on a community level.

### Other Soft Tissue Sarcoma Studies

Since soft tissue sarcomas are rare, the main epidemiological approach to investigating their etiology involves case-control studies. However, soft tissue sarcomas have been sought in several cohort studies. No excesses have been found associated with phenoxy herbicide spraying, but no conclusion can be drawn since the expected numbers are so low. For example, a study of 1926 Finnish herbicide sprayers who used 2,4-D and 2,4,5-T did not find any soft tissue sarcomas, but the expected number was only 0.1 (Riihimaki et al., 1982, 1983).

An excess of soft tissue sarcoma was noted among farmers, farm managers, and market gardeners in the United Kingdom (Balarajan and Acheson, 1984). There were 42 cases observed with a relative risk estimate of 1.7 (95% confidence interval 1.00–2.88). However, there was no excess among other agricultural workers, and no reason to link the excess to phenoxy herbicide exposure.

## *5.3 Malignant Lymphoma*

Rates of malignant lymphoma and multiple myeloma have been found to be elevated in agricultural occupations in a large number of studies (Sharp et al., 1986). Herbicides have been postulated as one of the number of possible agents which might explain these findings.

### Malignant Lymphoma and Phenoxyacetic Acid Herbicide Spraying

*Studies in Sweden.* As with soft tissue sarcoma, concern about phenoxy herbicides and lymphoma arose in Sweden when a number of male patients with histiocytic lymphoma reported past exposure to phenoxy herbicides. A case-control study of 169 cases of malignant lymphoma and 338 controls was then undertaken (Hardell et al., 1981; Hardell and Bengtsson, 1983). The study design, including ascertainment of exposure, was similar to the Swedish soft tissue sarcoma studies (Hardell and Sandstrom, 1979). A relative risk estimate of 4.8 (95% confidence interval 2.9–8.1) was obtained for exposure to phenoxy herbicides, excluding cases and controls exposed to chlorophenols. Stratifying by duration of exposure, the relative risk estimate was 4.3 for less than 90 days, and 7.0 for 90 days or more exposure to phenoxy herbicides. Most phenoxy herbicide exposed cases reported exposure to both 2,4,5-T and 2,4-D (27 cases) with 7 reporting exposure to 2,4-D alone, and 5 reporting exposure to MCPA alone (Hardell, 1981b).

A brief report from a clinical study of 123 male patients with non-Hodgkin's lymphoma (NHL) in Sweden found that 4 of 5 patients with cutaneous lesions reported spraying large areas with phenoxy acid herbicides, whereas only 7 of the remaining 118 patients with non-cutaneous lesions reported a similar occupational exposure to phenoxy herbicides ($p < 0.001$) (Olsson and Brandt, 1981).

The recent Swedish census-based study referred to in the section on soft tissue sarcoma also examined rates for non-Hodgkin's lymphoma in the agriculture and forestry cohort (Wiklund, 1986). The relative risk estimate for agricultural workers getting non-Hodgkin's lymphoma was 0.97, with 95% confidence interval of 0.89–1.06. This finding is in sharp contrast to the results of the Swedish case-control study. As was noted for the soft tissue sarcoma findings, the discrepancy between the studies cannot be dismissed because of lack of individual exposure data in the census-based study.

*Studies in New Zealand.* In New Zealand, an analysis of reported occupations appearing on the National Cancer Registry indicated an excess of malignant lymphoma and multiple myeloma among agricultural occupations (Pearce et al., 1985). The main findings for malignant lymphoma concerned ICD code 202 involving non-Hodgkin's lymphoma other than lymphosarcoma and reticulosarcoma; a relative risk estimate of 1.76 (95% confidence interval 1.03–3.02) was obtained for those under age 65 working in agriculture. However, a subsequent interview study did not suggest that exposure to phenoxy herbicides was the explanation, because a relative risk estimate of 1.3 (90% confidence interval 0.7–2.5) was obtained when using other cancers as controls and a relative risk of 1.0 (90% confidence interval 0.5–2.0) when using general population controls (Pearce et al., 1986).

*United States Case-Control Studies.* Some supportive evidence for an association of malignant lymphoma and exposure to phenoxy herbicides comes from a U.S. case-control study of non-Hodgkin's lymphoma (Hoar et al., 1986). The study involved 170 patients with non-Hodgkin's lymphoma who were interviewed by telephone. The relative risk estimate for farm herbicide exposure was 1.6 (95% confidence interval 0.9–2.6) based on 40 exposed cases. The estimate increased to 6.0 (95% confidence interval 1.9–19.5) when the focus was on men exposed to herbicides for more than 20 days per year. Frequent users who mixed or applied the herbicides themselves had a relative risk estimate of 8.0 (95% confidence interval 2.3–27.9). Surprisingly, there was no association with number of years of herbicide exposure after adjustment was made for days of herbicide exposure per year.

The predominant phenoxy herbicide used was 2,4-D, although the extent of individual use of 2,4-D by each study subjects was not requested. A study of 2,4-D contamination did not find any PCDDs or PCDFs for a detection limit of 0.01 to 0.05 ppm (μg/g) (Norstrom et al., 1979). However, a more recent analysis found 6.8 ppb (ng/g) of 2,3,7,8-TCDD in one sample of 2,4-D, measured with a detection limit of about 1 ppb (Hagenmaier, 1986). This is much lower than levels of contamination of 2,4,5-T with 2,3,7,8-TCDD. Thus, one cannot interpret this lymphoma study as providing evidence of carcinogenicity of PCDDs or PCDFs.

Another U.S. case-control study of non-Hodgkin's lymphoma produced mixed evidence (Woods et al., 1987). The study involved 576 cases of non-Hodgkin's lymphoma and 694 randomly selected controls. Among study subjects with occupations involving potential past exposure to phenoxy herbicides, the relative risk estimate was 1.07 (95% confidence interval 0.8–1.4). The estimates associated with exposure specifically to 2,4-D and 2,4,5-T were 0.73 (95% confidence interval 0.4–1.3) and 0.98 (95% confidence interval 0.5–2.0), respectively. However, the relative risk estimate was 1.71 (95% confidence interval 1.04–2.8) for potential exposure to phenoxy herbicides for 15 years or more during the period prior to 15 years before cancer diagnosis, and the estimate for forestry herbicide spraying was 4.8 (95% confidence interval 1.2–19.4). The authors noted that they did not find increased cancer risks for numerous other occupations and activities involving comparable opportunity for exposure, including those involved in other herbicide spraying activities or among those who worked as herbicide formulators or applicators per se.

### Malignant Lymphoma in the Chemical Manufacturing Industry

The chemical worker cohort involving 2192 workers manufacturing chlorophenols with potential for exposure to PCDDs referred to in the soft tissue sarcoma section sustained 5 deaths due to non-Hodgkin's lymphoma when 2.6 were expected (SMR = 192, 95% confidence interval 62–449) (Ott et al., 1987). There was no evidence for an increasing trend of risk with an increasing TCDD exposure intensity score, nor with a cumulative dose index.

The Danish cohort study of phenoxy herbicide manufacturers which reported an excess of soft tissue sarcomas did not find an excess for malignant lymphoma (Lynge, 1985). There were 7 cases found among the workers (relative risk = 1.30, 95% confidence interval 0.52–2.69). Moreover, 6 of the 7 cases were assigned to the department manufacturing pigments and not phenoxy herbicides.

### *5.4 Cancer of Other Sites*

A variety of studies have reported the occurrence of tumors other than soft tissue sarcoma and malignant lymphoma, without any consistent patterns emerging. Studies predominantly involve cohort studies of herbicide sprayers or chlorophenol manufacturing workers exposed to 2,3,7,8-TCDD.

The largest cohort study of phenoxy herbicide sprayers involved 1926 2,4-D and 2,4,5-T sprayers (Riihimaki et al., 1982). Between 1972 and 1980, there were 26 cancer deaths among the cohort with 36.5 expected (relative risk = 0.7, 95% confidence interval 0.5–1.0). Allowing for a 10-year latency, there were 20 cancers found with 24.3 expected (relative risk = 0.8, 95% confidence interval 0.5–1.3). Of these 20 cancers, 12 were lung cancers (11.1 expected) and 4 stomach or oesophageal cancers (3.7 expected). As noted previously, there were no cases of soft tissue sarcoma or malignant lymphoma. It should be noted that duration of exposure was low with few workers using phenoxy herbicides for more than 2 months.

A small cohort study of 142 forestry workers in Sweden who were exposed to 2,4,5,-T and 2,4-D did not find any overall increase in cancer mortality or cancer incidence (Hogstedt and Westerlung, 1980). There were 5 cancer deaths with 6.4 expected on the basis of national rates (relative risk = 0.8, 95% confidence interval 0.3–1.8). Based on tumor registry data, there were 8 incident cases of cancer with 9.8 expected (relative risk = 0.8, 95% confidence interval 0.4–1.6). However, 5 of the incident cases of cancer were among foremen with 1.4 expected (relative risk = 3.6, 95% confidence interval 1.2–8.3). Foremen had been exposed to phenoxy herbicides for an average of 176 days, compared with 30 days for other workers. Interpretation of this association must also consider the deficit among the other workers (3 cancers observed, 8.4 expected, relative risk = 0.4, 95% confidence interval 0.1–1.0). The eight incident cancer cases included two of the pancreas and two of the prostate. No more than one tumor was found for any other site.

Phenoxy herbicides appeared to double the risk of nasal and nasopharyngeal cancer in one study, although the finding was not statistically significant (95% confidence interval 0.9–4.7), and the main study findings concerned chlorophenol exposure (Hardell et al., 1982).

Excess cancers have been reported among Swedish railroad workers (Axelson and Sundell, 1974; Axelson et al., 1980). The initial report suggested an association with amitrol, but an update of the study suggested a weaker association with amitrol, but a stronger one with phenoxy herbicides. There were 17 tumors found with 11.85 expected. In the subcohort with combined exposure to amitrol and phenoxy herbicides there were 6 cases and 1.78 expected with a relative risk of 3.4 (95% confidence interval 1.2–7.3). No tumor site predominated. In the combined group there were 3 stomach cancers (1.34 expected). There were 2 stomach cancers in the subcohort with exposure to phenoxy acids alone (0.33 expected).

A study of agricultural workers in Germany (GDR) found an excess of lung cancer among pesticide exposed workers with rates said to be 20 times higher than population rates (Barthel, 1976). A subsequent larger cohort study by the same author found 169 neoplasms among the cohort of 1791 agricultural workers (Barthel, 1981). Of these cancers, 59 were bronchial which was about twice as many as expected based on the average GDR experience (relative risk =2.1, confidence interval 1.6–2.8). There was no increase for tumors of other sites. The average duration of exposure to pesticides among the bronchial carcinoma cases was 13.6 years, and the average latency from first exposure to diagnosis was 17.5 years. The smoking habits of the exposed group did not differ from those of the general population. The agents thought to be of most concern were arsenic-containing compounds, DDT, HCH, and phenoxy acids. The main phenoxy acids used were 2,4-D, MCPA, and 2,4,5-T.

There have been several cohort studies involving workers manufacturing trichlorophenols. While the cohorts involved were generally small, these studies are important because the exposure to PCDDs, in particular 2,3,7,8-TCDD, was much greater than was the case for phenoxy herbicide sprayers. Several of the studies involved workers who developed chloracne. Cases with soft tissue sarcoma have been reported for some of these cohorts, and this aspect has been dealt with in the earlier section on this topic.

A small study of 61 males involved in a 1974 U.S. trichlorophenol plant accident included 49 who developed chloracne. Follow-up to 1978 detected 3 cancer deaths with 1.6 expected (relative risk = 1.9, 95% confidence interval 0.4–5.5) (Cook et al., 1980). Follow-up of a similar accident in England had not detected any cancer cases after 10 years of follow-up (May, 1982). A similar 10-year follow-up of an accident in Czechoslovakia reported 2 cases of cancer, both lung cancer, among the 6 deaths observed (Pazderova-Vejlupkova et al., 1981). However, a longer 27-year follow-up of 74 workers exposed in a comparable accident in a plant in Germany reported 7 cancer deaths with 4.1 expected (relative risk = 1.7, 95% confidence interval 0.7–3.5) (Thiess et al., 1982). Three of the deaths were from stomach cancer, with 0.6 expected from regional mortality data (relative risk = 5, 95% confidence interval 1.0–14.6). A more recent publication gives further follow-up data for these workers (Lehnert and Szadkowski, 1986). A total of 17 cancer deaths had occurred up to 1983, out of a total of 43 deaths. There were no additional deaths from stomach cancer. While data concerning expected numbers were not presented, the total number of cancer deaths and the deaths due to stomach cancer were reported to be greater than expected. However, these differences were reported not to be statistically significant.

An accident which occurred in 1949 in the Nitro, West Virginia, plant has already been mentioned in the context of chloracne (Zack and Suskind, 1980). Follow-up of 121 workers who developed chloracne identified 9 deaths from cancer with 9.04 expected (relative risk = 1.0, 95% confidence interval 0.5–1.9) (Zack and Suskind, 1980). One was a soft tissue sarcoma case included in the soft tissue sarcoma review section. There were 4 cases of lung cancer, but no cases of stomach cancer. A more recent study of workers at this plant demonstrated an increased risk of bladder cancer with 11 cases identified, but all had been exposed to 4-aminobiphenyl, a proven bladder carcinogen (Moses et al., 1984).

The relatively large cohort of 2192 chemical workers in a U.S. manufacturing plant who had potential exposure to PCDDs identified 81 cancer deaths with 79.3 expected (relative risk = 1.02, 95% confidence interval 0.8–1.3) (Ott et al., 1987). The excess of non-Hodgkin's lymphoma has already been discussed. There was also a small excess of stomach cancer deaths (6 observed vs 3.8 expected, relative risk = 1.6, 95% confidence interval 0.6–3.4), but no increase in lung cancer deaths (23 observed vs 27.9 expected, relative risk = 0.8, 95% confidence interval 0.5–1.2). There were no increases in stomach cancer risk with increasing intensity score for 2,3,7,8-TCDD exposure.

Finally, a follow-up study of 80 workers in a trichlorophenol plant, most of whom developed chloracne, detected 2 deaths from cancer, both lung cancers, among 6 deaths in the 10-year follow-up period (Pazderova-Vejlupkova et al., 1981).

### *5.5 Interpretation of Human Cancer Findings*

The Swedish case-control studies have reported excess soft tissue sarcoma and malignant lymphoma with very short duration of exposure to 2,4,5-T, but these findings have not been confirmed in other countries, and recent Swedish evidence does not support them.

Workers who have developed chloracne in trichlorophenol plants have presumably been exposed to much higher doses of 2,3,7,8-TCDD than pesticide applicators who have only sprayed phenoxy herbicides for a few months, but studies of trichlorophenol workers do not provide convincing evidence of increased cancer risk.

Although 2,3,7,8-TCDD is a very potent animal carcinogen, human exposure via phenoxy herbicides may have been too low to elicit effects. Levels of around 1 ppm of 2,3,7,8-TCDD were probably common in older formulations of 2,4,5-T (Norstrom et al., 1979; Smith and Pearce, 1986). A review of studies of 2,4,5-T applicators suggested that the maximum absorbed dose of 2,4,5-T was not likely to exceed 0.1 mg/kg of body weight per day (Leng et al., 1982). One applicator who had a leaking back pack was exposed to 0.2 mg/kg/day. If one assumes an average 1 ppm level of 2,3,7,8-TCDD for 2,4,5-T used prior to 1970, then the daily dose to an applicator would be 0.0002 μg/kg/day of 2,3,7,8-TCDD if 0.2 mg/kg/day of 2,4,5-T were absorbed, and if the dioxin was absorbed in the same manner as 2,4,5-T.

Liver tumors have been produced in experimental animals with doses of 2,3,7,8-TCDD as low as .01 μg/kg/day for 2 years, the lowest observable effect level (LOEL) for cancer. A herbicide applicator spraying 2,4,5-T every day would only be receiving 1/50 of the LOEL daily dose for the experimental animals. Since spraying commenced in the late 1940s, and since 2,3,7,8-TCDD levels in 2,4,5-T reduced dramatically in the 1970s, the longest potential exposure at significant levels is about a third of a 70-year lifetime, and might involve spraying on 3 days per week. This exposure would correspond to $\frac{1}{50} \times \frac{1}{3} \times \frac{3}{7} = \frac{1}{350}$ of the animal LOEL for cumulative lifetime exposure. However, very few applicators have worked this long. In fact, applicators reported in Swedish studies, who mostly had exposures totaling less than 1 work year, would have experienced a cumulative lifetime exposure of the order of $\frac{1}{8000}$ (i.e., $\frac{1}{50} \times \frac{3}{7} \times \frac{1}{70}$) that of the animal LOEL. Even correcting this figure for pharmacokinetic differences between animals and humans, one would not expect to detect increased cancer risks in studies of sprayers assuming that their sensitivity to carcinogenic effects of 2,3,7,8-TCDD is comparable to experimental animals.

As has been suggested in previous reviews, further research is needed (Coggon and Acheson, 1982; Axelson, 1984; Ayres et al., 1985; Fingerhut et al., 1987). Although one reviewer has concluded that the available data indicate an increased risk for cancer from phenoxy herbicide exposure (Sterling and Arundel, 1986b), at this stage there is no consistent evidence that 2,3,7,8-TCDD is a human carcinogen at the levels of exposure involved. In addition, the evidence for carcinogenicity of the phenoxy herbicides themselves must be counterbalanced by the absence of supportive animal data.

## 6 Summary and Conclusions

Causal inference in epidemiology is based on a number of different criteria outlined in the introduction. These include: the likelihood that findings are not attributable to chance; the consistency of findings from study to study; the strength of the findings, including strength of any dose-response relationships; and the biological plausibility of the reported associations.

The only association between exposure to PCDDs and a health effect which satisfactorily meets these criteria is that between high level, short term exposure to 2,3,7,8-TCDD, and chloracne. Various other effects have been found which are possibly due to PCDDs including porphyria and transient changes in liver function tests. It is also possible that PCDDs may have caused gastrointestinal ulcers, peripheral neuropathy, and changes in blood lipid levels. Evidence for immune system effects is questionable.

Various toxic effects of PCDFs, including swelling and hypertrophy of the Meibomian glands and chloracne, have been seen in patients with Yusho and Yu-Cheng disease.

Studies concerning reproductive effects, taken overall, are consistent in not finding effects of PCDDs among men experiencing occupational exposure. Sufficient data on women workers are lacking. Community exposures to dioxins have been lower than worker exposures and have not provided consistent evidence of reproductive effects.

The results of human cancer studies of exposure to PCDDs are inconsistent. Some studies find increased risks of soft tissue sarcoma; but most do not. There is some evidence for increased malignant lymphoma incidence with exposure to phenoxy herbicides; but it is unlikely that the effects, if real, are attributable to PCDDs. Extrapolating from animal data, the levels of contamination of phenoxy herbicides with 2,3,7,8-TCDD are below amounts which would be expected to result in detectable increases in cancer risks. Some industrial workers have experienced higher exposures, but no consistent findings have emerged. Taken collectively, the available human studies do not demonstrate an increased risk of cancer at the levels of exposure experienced.

## 7 References


Axelson O (1980) A note on observational bias in case-referent studies in occupational health epidemiology [letter]. Scand. J. Work Environ. Health 6: 80–82

Axelson O (1984) The health effects of phenoxy acid herbicides. In: Herrington JM (ed) Recent advances in occupational health. Churchill Livingstone, Edinburgh, pp 253–266

Axelson O, Sundell L (1974) Herbicide exposure, mortality and tumor incidence: an epidemiological investigation on Swedish railroad workers. Work Environ. Health 11: 21–28

Axelson O, Sundell L, Andersson K, Edling C, Hogstedt C, Kling H (1980) Herbicide exposure and tumor mortality: an updated epidemiologic investigation on Swedish railroad workers. Scand. J. Work Environ. Health 6: 73–79

Ayres SM, Webb KB, Evans RG, Mikes J (1985) Is 2,3,7,8-TCDD (dioxin) a carcinogen for humans? Environ. Health Perspect. 62: 329–335

Balarajan R, Acheson ED (1984) Soft tissue sarcomas in agriculture and forestry workers. J. Epidemiol. Community Health 38: 113–116

Barthel E (1976) [High incidence of lung cancer in persons with chronic professional exposure to pesticides in agriculture]. Z. Erk. Atmungsorgane 146: 266–274

Barthel E (1981) Increased risk of lung cancer in pesticide-exposed male agricultural workers. J. Toxicol. Environ. Health 8: 1027–1040

Bleiberg J, Wallen M, Brodkin R, Applebaum IL (1964) Industrially acquired porphyria. Arch. Dermatol. 89: 793–797

Bond GG, Ott MG, Brenner FE, Cook RR (1982) Medical and morbidity surveillance findings among employees potentially exposed to TCDD. Br. J. Ind. Med. 40: 318–324

Bruzzi P (1983) Birth defects in the TCDD polluted area of Seveso: results of a four-year follow-up. In: Coulston F, Pocchiari F (eds) Accidental exposure to dioxins: human health aspects. Academic, New York, pp 271–280

Caramaschi F, Corno G, Favaretti C, Giambelluca SE, Montesarchio E, Fara GM (1981) Chloracne following environmental contamination by TCDD in Seveso, Italy. Int. J. Epidemiol. 10: 135–143

Chen PH, Wong CK, Rappe C, Nygren M (1985) Polychlorinated biphenyls, dibenzofurans and quaterphenyls in toxic rice-bran oil and in the blood and tissues of patients with PCB poisoning (Yu-Cheng) in Taiwan. Environ. Health Perspect. 59: 59–65

Chen PH, Hsu ST (1987) PCB poisoning from toxic rice-bran oil in Taiwan. In: Waid JS (ed) PCBs and the environment, vol 3. CRC, Boca Raton, FL, pp 27–38

Coggon D, Acheson ED (1982) Do phenoxy herbicides cause cancer in man? Lancet 1: 1057–1059

Colton T (1986) Herbicide exposure and cancer. J. Am. Med. Assoc. 256: 1176–1178

Constable JD, Hatch MC (1985) Reproductive effects of herbicide exposure in Vietnam: recent studies by the Vietnamese and others. Teratogen. Carcinogen. Mutagen. 5: 231–250

Cook RR, Townsend JC, Ott MG, Silverstein LG (1980) Mortality experience of employees exposed to 2,3,7,8-tetrachlorodibenzo-*p*-dioxin (TCDD). J. Occup. Med. 22: 530–532

Coulston F, Olajos EJ (1980) Panel report: Panel to discuss the epidemiology of 2,4,5-T. Ecotoxicol. Environ. Safety 4: 96–102

Crow KD (1983) Significance of cutaneous lesions in the symptomatology of exposure to dioxins and other chloracnegens. In: Tucker RE (ed) Human and environmental risks of chlorinated dioxins and related compounds. Plenum, New York, pp 605–612

Dan BB (1984) Vietnam and birth defects. J. Am. Med. Assoc. 252: 936–937

Donovan JW, MacLennan R, Adena M (1984) Vietnam service and the risk of congenital anomalies: a case-control study. Med. J. Aust. 140: 394–397

Doss M, Sauer H, von Tiepermann R, Colombi AM (1984) Development of chronic hepatic porphyria (porphyria cutanea tarda) with inherited uroporphyrinogen decarboxylase deficiency under exposure to dioxin. Int. J. Biochem. 16: 369–373

Erickson JD, Mulinare J, McClain PW, Fitch TG, James LM, McClearn AB, Adams MJ Jr (1984) Vietnam veterans' risk for fathering babies with birth defects. J. Am. Med. Assoc. 252: 903–912

Eriksson M, Hardell L, Berg NO, Moller T, Axelson O (1981) Soft-tissue sarcomas and exposure to chemical substances: a case-referent study. Br. J. Ind. Med. 38: 27–33

Evans RG, Webb KB, Knutsen AP, Roodman ST, Roberts D, Bagby J, Garrett WA, Andrews JS Sr (1987) A medical follow-up of the health effects of long-term exposure to 2,3,7,8-tetrachlorodibenzo-*p*-dioxin. Presented at Dioxin '87: Seventh International Symposium on Chlorinated Dioxins and Related Compounds, October 4–9, Las Vegas. [Abstract]. Available from: [University of Nevada, Las Vegas]

Field B, Kerr C (1979) Herbicide use and incidence of neural-tube defects. Lancet 1: 1341–1342

Filippini G, Bordo B, Crenna P, Massetto N, Musicco M, Boeri R (1981) Relationship between clinical and electrophysiological findings and indicators of heavy exposure to 2,3,7,8-tetrachlorodibenzo-dioxin. Scand. J. Work Environ. Health 7: 257–262

Fingerhut MA, Halperin WE, Honchar PA, Smith AB, Groth DH and Russell WO (1984) An evaluation of reports of dioxin exposure and soft tissue sarcoma pathology in U.S. chemical workers. In: Poland A and Kimbrough RD (eds) Banbury Report 18. Biological mechanisms of dioxin action. Cold Spring Harbor, NY. Cold Spring Harbor Laboratory. pp 461–470

Fingerhut MA, Sweeney MH, Halperin WE and Schnorr TM (1987) Epidemiology of populations exposed to dioxins. In: Exner JH (ed) Solving hazardous waste problems: learning from dioxins. ACS Symposium Series 338. American Chemical Society, Washington, DC, pp 143–161

Fox JP, Hall CE, Elveback LR (1970) Epidemiology, man and disease. Macmillan, New York, p. 185

Friedman JM (1984) Does Agent Orange cause birth defects? Teratology 29: 193–221

Greenwald P, Kovasznay B, Collins DN, Therriault G (1984) Sarcomas of soft tissues after Vietnam service. J. Natl. Cancer Inst. 73: 1107–1109

Gross ML, Lay JO, Lyon PA, Lippstreu D, Kangas N, Harless RL, Taylor SE, Dupuy AE Jr (1984) 2,3,7,8-Tetrachlorodibenzo-*p*-dioxin levels in adipose tissue of Vietnam veterans. Environ. Res. 33: 261–268

Hagenmaier H (1986) Determination of 2,3,7,8-tetrachlorodibenzo-*p*-dioxin commercial chlorophenols and related products. Fresenius Z. Anal. Chem. 325: 603–606

Hanify JA, Metcalf P, Nobbs CL, Worsley KJ (1981a) Aerial spraying of 2,4,5-T and human birth malformations. Final report of an epidemiological study carried out in the Northland region of New Zealand. Auckland, New Zealand: Northland Birth Surveys Final Report. 27 p. Available from: Northland Births Survey, Box 6256, Aukland 1, New Zealand

Hanify JA, Metcalf P, Nobbs CL, Worsley KJ (1981b) Aerial spraying of 2,4,5-T and human birth malformations: an epidemiological investigation. Science 212: 349–351

Hardell L (1981a) Relation of soft-tissue sarcoma, malignant lymphoma and colon cancer to phenoxy acids, chlorophenols and other agents. Scand. J. Work Environ. Health 7: 119–130

Hardell L (1981b) Epidemiological studies on soft-tissue sarcoma and malignant lymphoma and their relation to phenoxy acid or chlorophenol exposure. Umea University Medical Dissertations, New Series No. 65. University of Umea, Sweden. 139 p

Hardell L, Bengtsson NO (1983) Epidemiological study of socioeconomic factors and clinical findings in Hodgkin's disease and reanalysis of previous data regarding chemical exposure. Br. J. Cancer 48: 217–225

Hardell L, Eriksson M (1986) Soft-tissue sarcoma and exposure to dioxins. Lancet 2: 868

Hardell L, Sandstrom A (1979) Case-control study: soft-tissue sarcomas and exposure to phenoxyacetic acids or chlorophenols. Br. J. Cancer 39: 711–717

Hardell L, Eriksson M, Lenner P, Lundgren E (1981) Malignant lymphoma and exposure to chemicals, especially organic solvents, chlorophenols and phenoxy acids: a case-control study. Br. J. Cancer 43: 169–176

Hardell L, Johansson B, Axelson O (1982) Epidemiological study of nasal and nasopharyngeal cancer and their relation to phenoxyacid or chlorophenol exposure. Am. J. Ind. Med. 3: 247–257

Hatch MC (1984) Reproductive effects of the dioxins. In: Lowrance WW (ed) Public health risks of the dioxins. The Rockefeller University, New York, pp 255–274

Hoar SK, Blair A, Holmes FF, Boysen CD, Robel RJ, Hoover R, Fraumeni JF Jr (1986) Agricultural herbicide use and risk of lymphoma and soft-tissue sarcoma. J. Am. Med. Assoc. 256: 1141–1147

Hobson LB (1984) Human effects of TCDD exposure. Bull. Environ. Contam. Toxicol. 33: 696–701

Hoffman RE, Stehr-Green PA, Webb KB, Evans RG, Knutsen AP, Schramm WF, Staake JL, Gibson BB, Steinberg KK (1986) Health effects of long-term exposure to 2,3,7,8-TCDD. J. Am. Med. Assoc. 255: 2031–2038

Hogstedt C, Westerlung B (1980) [Survey of the death causes of forest workers with and without exposure to phenoxyacid chemicals]. Lakartinoningen 77: 1828–1831

Honchar PA, Halperin WE (1981) 2,4,5-T, trichlorophenol, and soft-tissue sarcoma. Lancet 1: 268–269

Ideo G (1984) Final report on the results of D-glucaric acid determinations done at Seveso in adults and children in the period 1978–1982. Report to the Special Office for Seveso, Italy. 7 p. Available from: The Office

Japan-United States (1985) Joint Seminar on Toxicity of Chlorinated Biphenyls, Dibenzofurans, Dibenzodioxins, and Related Compounds. April 25–28, 1983, Fukuoka, Japan. Environ. Health Perspect. 59: 1–181

Jirasek L, Kalensky J, Kubec K (1973) [Acne chlorina and porphyria cutanea tarda during the manufacture of herbicides]. Cesk. Dermatol. 48: 306–317

Jirasek L, Kalensky J, Kubec K, Pazderova J, Lukas E (1974) [Acne chlorina, porphyria cutanea tarda and other manifestations of general poisoning during the manufacture of herbicides]. II. Cesk. Dermatol. 49: 145–157

Kang HK, Weatherbee L, Breslin PP, Lee Y, Shepard BM (1986) Soft tissue sarcomas and military service in Vietnam: a case comparison group analysis of hospital patients. J. Occup. Med. 28: 1215–1218

Kang H, Enziger F, Breslin P, Feil M, Lee Y, Shepard B (1987) Soft tissue sarcoma and military service in Vietnam: a case-control study. J. Natl. Cancer Inst. 79: 693–699

Kashimoto T, Miyata H (1987) Differences between Yusho and other kinds of poisoning involving only PCBs. In: Waid JS (ed) PCBs and the environment. CRC, Boca Raton, FL, pp 1–26

Kashimoto T, Miyata H, Kunita S, Tung T, Hsu ST, Chang KJ, Tang SY, Ohi G, Nakagawa J, Yamamoto SI (1981) Role of polychlorinated dibenzofuran in Yusho (PCB poisoning). Arch. Environ. Health 36: 321–326

Knutsen AP (1984) Immunologic effects of TCDD exposure in humans. Bull. Environ. Contam. Toxicol. 33: 673–681

Knutsen AP, Roodman ST, Evans RG, Mueller KR, Webb KB, Green PS, Hoffman RE, Schramm WF (1987) Immune studies in dioxin-exposed Missouri residents: Quail Run. Bull. Environ. Contam. Toxicol. 39: 481–489

Kociba RJ, Keeler PA, Park CN, Gehring PJ (1976) 2,3,7,8-Tetrachlorodibenzo-*p*-dioxin (TCDD): results of a 13-week oral toxicity study in rats. Toxicol. Appl. Pharmacol. 35: 553–574

Kociba RJ, Keyes DG, Beyer JE, Carreon RM, Wade CE, Dittenber DA, Kalnins RP, Franson LE, Park CN, Bernard SD, Hummel RA, Humiston CG (1978) Results of a two-year chronic toxicity and oncogenicity study of 2,3,7,8-tetrachlorodibenzo-*p*-dioxin in rats. Toxicol. Appl. Pharmacol. 46: 279–303

Kogan MD, Clapp RW (1985) Mortality among Vietnam veterans in Massachusetts, 1972–1983. The Commonwealth of Massachusetts: Office of the Commissioner of Veterans' Services

Kunita N, Hori S, Obana H, Otake T, Nishimura H, Kashimoto T, Ikegami N (1985) Biological effect of PCBs, PCQs and PCDFs present in the oil causing Yusho and Yu-Cheng. Environ. Health Perspect. 59: 79–84

Kuratsune M, Yoshimura T, Matsuzaka J, Yamaguchi A (1972) Epidemiologic study on Yusho, a poisoning caused by ingestion of rice oil contaminated with a commercial brand of polychlorinated biphenyls. Environ. Health Perspect. 1: 119–128

Lathrop GD, Wolfe WH, Albanese RA, Moynahan PM (1984) An epidemiological investigation of health effects in Air Force personnel following exposure to herbicides: baseline morbidity study results. USAF School of Aerospace Medicine. Brooks Air Force Base, TX

Lathrop GD, Wolfe WH, Machado SG, Michalek JE, Karrison TG, Miner JC, Grubbs WD, Peterson MR, Thomas WF (1987) An epidemiologic investigation of health effects in Air Force personnel following exposure to herbicides: first follow-up examination results. vol 1. Available from: National Technical Information Service, Springfield, VA

Lawrence CE, Reilly AA, Quickenton P, Greenwald P, Page WF, Kuntz AJ (1985) Mortality patterns of New York State Vietnam veterans. Am. J. Public Health 75: 277–279

Lehnert VG, Szadkowski D (1986) [The carcinogenicity of 2,3,7,8-tetrachlorodibenzo-*p*-dioxin in humans: evaluation of liability]. Arbeitsmed. Sozialmed. Praventivmed. 21: 153–175

Leng ML, Lavy TL, Ramsey JC, Braun WH (1982) Review of studies with 2,4,5-T in humans including applicators under field conditions. In: Plimmer JR (ed) Pesticide residues and exposure. vol 182. American Chemical Society, Washington, DC, pp 133–156

Lynge E (1985) A follow-up study of cancer incidence among workers in manufacture of phenoxy herbicides in Denmark. Br. J. Cancer 52: 259–270

Martin JV (1984) Lipid abnormalities in workers exposed to dioxin. Br. J. Ind. Med. 41: 254–256

Mastroiacova P, Spagnolo A, Marni E, Meazza L, Bertollini R, Segni G (1988) Birth defects in the Seveso area after TCDD contamination. J. Am. Med. Assoc. 259: 1668–1672

May G (1982) Tetrachlorodibenzodioxin: a survey of subjects ten years after exposure. Br. J. Ind. Med. 39: 128–135

Merlo F and Puntoni R (1986) Soft-tissue sarcomas, malignant lymphomas, and 2,3,7,8-TCDD exposure in Seveso. Lancet 2: 1455

Mocarelli P, Marocchi A, Brambilla P, Gerthoux PM, Young DS, Mantel N (1986) Clinical laboratory manifestations of exposure to dioxins in children: a six-year study of the effects of an environmental disaster near Seveso, Italy. J. Am. Med. Assoc. 256: 2687–2695

Moses M, Lilis R, Crow KD, Thornton J, Fischbein A, Anderson HA, Selikoff IJ (1984) Health status of workers with past exposure to 2,3,7,8-tetrachlorodibenzo-*p*-dioxin in the manufacture of 2,4,5-trichlorophenoxyacetic acid: comparison of findings with and without chloracne. Am. J. Ind. Med. 5: 161–182

Nelson CJ, Holson JF, Green HG, Gaylor DW (1979) Retrospective study of the relationship between agricultural use of 2,4,5-T and cleft palate occurrence in Arkansas. Teratology 19: 377–383

Norstrom A, Rappe C, Lindahl R, Buser HR (1979) Analysis of some older Scandinavian formulations of 2,4-dichlorophenoxyacetic acid and 2,4,5-trichloro-phenoxyacetic acid for contents of chlorinated dibenzo-*p*-dioxins and dibenzofurans. Scand. J. Work Environ. Health 5: 375–378

Olsson H, Brandt L (1981) Non-Hodgkin's lymphoma of the skin and occupational exposure to herbicides. Lancet 2: 579

Ott MG, Olson RA, Cook RR, Bond GG (1987) Cohort mortality study of chemical workers with potential exposure to the higher chlorinated dioxins. J. Occup. Med. 29: 422–429

Pazderova-Vejlupkova J, Nemcova M, Pickova J, Jirasek L, Lucas E (1981) The development and prognosis of chronic intoxication by tetrachlorodibenzo-*p*-dioxin in men. Arch. Environ. Health 36: 5–11

Pearce NE, Smith AH, Fisher DO (1985) Malignant lymphoma and multiple myeloma linked with agricultural occupations in a New Zealand cancer registry-based study. Am. J. Epidemiol. 121: 225–237

Pearce NE, Smith AH, Howard JK, Sheppard RA, Giles HJ, Teague CA (1986) Non-Hodgkin's lymphoma and exposure to phenoxyherbicides, chlorophenols, fencing work, and meat works employment: a case-control study. Br. J. Ind. Med. 43: 75–83

Poland AP, Smith D, Metter G, Possick P (1971) A health survey of workers in a 2,4,-D and 2,4,5,-T plant with special attention to chloracne, porphyria cutanea tarda, and psychological parameters. Arch. Environ. Health 22: 316–327

Puntoni R, Merlo F, Fini A, Meazza L, Santi L (1986) Soft tissue sarcomas in Seveso. Lancet. 2: 525

Reggiani G (1983) Anatomy of a TCDD spill: the Seveso accident. Hazard Assess. Chem. Curr. Dev. 2: 269–342

Riihimaki V, Asp S, Hernberg S (1982) Mortality of 2,4-dichlorophenoxyacetic acid and 2,4,5-trichlorophenoxyacetic acid herbicide applicators in Finland. Scand. J. Environ. Health 8: 37–42

Riihimaki V, Asp S, Pukkala E, Hernberg S (1983) Mortality and cancer morbidity among chlorinated phenoxyacid applicators in Finland. Chemosphere 12: 779–784

Sharp DS, Eskenazi B, Harrison R, Callas P, Smith AH (1986) Delayed health hazards of pesticide exposure. Annu. Rev. Public Health 7: 441–471

Smith AH (1983) Problems in dose response interpretation in occupational epidemiology. J. Univ. Occup. Environ. Health 5 (Suppl.): 189–195

Smith AH, Pearce NE (1986) Update on soft tissue sarcoma and phenoxy herbicides in New Zealand. Chemosphere 15: 1795–1798

Smith AH, Matheson DP, Fisher DO, Chapman CJ (1981) Preliminary report of reproductive outcomes among pesticide applicators using 2,4,5-T. New Zealand Med. J. 93: 177–179

Smith AH, Fisher DO, Pearce NE, Chapman CJ (1982a) Congenital defects and miscarriages among New Zealand 2,4,5-T sprayers. Arch. Environ. Health 37: 197–200

Smith AH, Fisher DO, Pearce NE, Teague CA (1982b) Do agricultural chemicals cause soft tissue sarcoma? Initial findings of a case-control study in New Zealand. Community Health Studies 6: 114–119

Smith AH, Fisher DO, Giles HJ, Pearce N (1983) The New Zealand soft tissue sarcoma case-control study: interview findings concerning phenoxyacetic acid exposure. Chemosphere 12: 565–571

Smith AH, Pearce NE, Fisher DO, Giles HJ, Teague CA, Howard JK (1984) Soft tissue sarcoma and exposure to phenoxyherbicides and chlorophenols in New Zealand. J. Natl. Cancer Inst. 73: 1111–1117

Stehr PA, Stein G, Falk H, Sampson E, Smith SJ, Steinberg K, Webb K, Ayres S, Schramm W, Donnell HD, Gedney WB (1986) A pilot epidemiologic study of possible health effects associated with 2,3,7,8-tetrachlorodibenzo-*p*-dioxin contaminations in Missouri. Arch. Environ. Health 41: 16–22

Sterling TD, Arundel A (1986a) Review of recent Vietnamese studies on the carcinogenic and teratogenic effects of phenoxy herbicide exposure. Int. J. Health Service 16: 265–278

Sterling TD, Arundel AV (1986b) Health effects of phenoxy herbicides: a review. Scand. J. Work Environ. Health 12: 161–173

Strik JJTWA, Janssen MMT, Colombi AM (1980) The incidence of chronic hepatic porphyria in an Italian family. Am. J. Biochem. 12: 879–881

Suskind RR (1983) Long-term health effects of exposure to 2,4,5-T and/or its contaminants. Chemosphere 12: 769

Suskind RR (1985) Chloracne, “the hallmark of dioxin intoxication”. Scand. J. Work Environ Health 11: 165–171

Suskind RR, Hertzberg VS (1984) Human health effects of 2,4,5-T and its toxic contaminants. J. Am. Med. Assoc. 251: 2372–2380

Sweeney G, Barford D, Rowley B, Goddard G (1984) Mechanisms underlying the hepatotoxicity of 2,3,7,8-tetrachlorodibenzo-*p*-dioxin. In: Poland A and Kimbrough RD (eds) Banbury Report 18. Biological mechanisms of dioxin action. Cold Spring Harbor, NY: Cold Spring Harbor Laboratory, pp 225–239

Tenchini ML, Crimaudo C, Pacchetti G, Mottura A, Agosti S, De Carli L (1983) A comparative cytogenetic study on cases of induced abortions in TCDD-exposed and nonexposed women. Environ. Mutagen. 5: 73–85

Thiess AM, Frentzel-Beyme R, Link R (1982) Mortality study of persons exposed to dioxin in a trichlorophenol-process accident that occurred in the BASF AG on November 17, 1953. Am. J. Ind. Med. 3: 179–189

Thomas HF (1980) 2,4,5-T use and congenital malformation rates in Hungary. Lancet 2: 214–215

Tindall JP (1985) Chloracne and chloracnegens. J. Am. Acad. Dermatol. 13: 539–558

Townsend JC, Bodner KM, VanPeenen PFD, Olsen RD, Cook RR (1982) Survey of reproductive events of wives of employees exposed to chlorinated dioxins. Am. J. Epidemiol. 115: 695–713

U.S. Environmental Protection Agency (1979) Report of assessment of a field investigation of six-year spontaneous abortion rates in three Oregon areas in relation to forest 2,4,5-T spray practices. Alsea II Report. Washington, DC. Available from: National Technical Information Service, Springfield, VA

Vineis P, Terracini B, Ciccone G, Cignetti A, Colombo E, Donna A, Maffi L, Pisa R, Ricci P, Zanini E, Comba P (1986) Phenoxy herbicides and soft-tissue sarcomas in female rice weeders: a population-based case-referent study. Scand. J. Work Environ. Health 13: 9–17

Webb K, Ayres S, Slavin R, Knutsen A, Roodman S, Gedney WB, Schramm W, Hotchkiss RL, Miller R, Donnell HD (1984) Results of a pilot study of health effects due to 2,3,7,8-tetrachlorodibenzo-*p*-dioxin contamination-Missouri. J. Am. Med. Assoc. 251: 1139–1140

West Virginia. State Health Department (1986) West Virginia Vietnam-era veterans mortality study. West Virginia Residents. 1968–1983. Preliminary Report, January. 30 p. Available from: The Department

Wiklund K (1986) Cancer risks among agricultural workers in Sweden: cohort studies based on the

Swedish cancer-environment register. Department of Cancer Epidemiology, Radiumhemmet, Karolinska Hospital, Stockholm, Sweden. Available from: The Department, S-104 01, Stockholm, Sweden

Wiklund K, Holm LE (1986) Soft tissue sarcoma risk in Swedish agricultural and forestry workers. J. Natl. Cancer Inst. 76: 229–234

Woods JS, Polissar L, Severson RK, Heuser LS, Kulander BG (1987) Soft tissue sarcoma and non-Hodgkin's lymphoma in relation to phenoxyherbicide and chlorinated phenol exposure in western Washington. J. Natl. Cancer. Inst. 78: 899–910

Zack JA, Suskind RR (1980) The mortality experience of workers exposed to tetrachlorodibenzodioxin in a trichlorophenol process accident. J. Occup. Med. 22: 11–14

# 2.2 Mechanism of Action

**Mechanisms Working Group**[1]

**S.H. Safe, T. Gasiewicz and J.P. Whitlock Jr.**

1 Introduction . . . 61
1.1 Biologic and Toxic Effects . . . 61
1.2 Possible Initial Mechanisms of Action . . . 63
1.3 Receptor Model of Initial Toxic Mechanism . . . 65
2 Molecular Properties and Binding Characteristics of the Ah Receptor . . . 66
2.1 Molecular Properties of Ah Receptor-Ligand Complexes . . . 66
2.2 Ah Receptor: Radioligand Binding Affinities, Species Differences, and Distribution . . . 69
2.3 Stereoselective Ah Receptor-Ligand Interactions . . . 72
2.4 Ah Receptor: Ontogeny, Regulation, and Relationships Between Receptor Levels and Biological Response . . . 74
3 Mechanism of Action of 2,3,7,8,-TCDD and Related Compounds: Role of the Ah Receptor . . . 75
3.1 PCDDs and PCDFs: Structure-Induction Relationships . . . 75
3.2 PCDDs and PCDFs: Structure-Toxicity Relationships . . . 78
3.3 Genetic Studies . . . 80
3.4 Models for Studying Ah Receptor-Mediated Responses . . . 80
4 Conclusions . . . 83
5 References . . . 84

## 1 Introduction

### *1.1 Biologic and Toxic Effects*

The polychlorinated dibenzo-*p*-dioxins (PCDDs) and dibenzofurans (PCDFs) elicit a broad range of toxic and biologic effects in diverse animal species and mammalian cells in culture (Poland et al., 1979; Poland and Knutson, 1982; Safe, 1986; Whitlock, 1987). These effects include immunologic alterations, morphologic changes in epithelial tissues, tumor promotion, and a variety of biochemical changes such as alterations in enzyme activities, hormone concentrations, and receptor levels (Table 1). In many cases it is not yet known whether

[1] Ad Hoc Panel on Health Aspects of Polychlorinated Dibenzo-*p*-dioxins and Polychlorinated Dibenzofurans, Universities Associated for Research and Education in Pathology, Inc., Bethesda, Maryland, USA

Environmental Toxin Series, Vol. 3

**Table 1.** Biologic and toxic effects of PCDDs and PCDFs

| Response | Reference |
|---|---|
| Acute lethality | Kociba and Schwetz, 1982a,b |
| Wasting syndrome | Poland and Glover, 1980; Rozman et al., 1985 |
| Thymic involution and immunotoxicity | Poland and Glover, 1980; Vecchi et al., 1980, 1983; Clark et al., 1983; Nagarkatti et al., 1984 |
| Dermal toxicity | Knutson and Poland, 1982; Poland et al., 1982, 1984 |
| Reproductive toxicity and teratogenicity | Poland and Glover, 1980; Hassoun et al., 1984; Weber et al., 1984, 1985; Birnbaum et al., 1985, 1987 |
| Chick edema | Flick et al., 1973 |
| Porphyria | Jones and Sweeney, 1980 |
| Enzyme induction: | |
| Cytochrome P-450 and related monooxygenases | Whitlock, 1986, 1987; Nebert and Gonzalez, 1987 |
| DT diaphorase | Beatty and Neal, 1976; Kumaki et al., 1977 |
| Aldehyde dehydrogenase | Deitrich et al., 1977 |
| Epidermal transglutaminase | Puhvel et al., 1984; Puhvel and Sakamoto, 1987 |
| Glutathione S-transferase | Kirsch et al., 1975 |
| δ-Aminolevulinic acid synthetase | Poland and Glover, 1973 |
| UDP glucuronosyl transferase | Owens, 1977; Thunberg et al., 1984 |
| Modulation of receptor binding levels | Karenlampi et al., 1983; Hudson et al., 1985; Osborne and Greenlee, 1985 |
| EGF receptor | Madhukar et al., 1984 |
| Glucocorticoid receptor | Ryan et al., 1987 |
| Estrogen receptor | Romkes et al., 1987a; Romkes and Safe, 1988 |
| Progesterone receptor | Romkes and Safe, 1988 |
| Ah receptor | Sloop and Lucier, 1987 |
| Modulation of thyroid hormone levels | Bastomsky, 1977; Gupta et al., 1983; Potter et al., 1983, 1986; Pazdernik and Rozman, 1985; Rozman et al., 1985; Lamb et al., 1986; Henry and Gasiewicz, 1987; Kelling et al., 1987 |
| Modulation of steroid metabolism enzymes | Moore et al., 1985, 1987; Mebus et al., 1987 |
| Depletion of vitamin A levels | Thunberg et al., 1979 |

a particular effect represents a primary or secondary response to the compounds. Nevertheless, a mechanism that seeks to explain the biological effects of PCDDs and PCDFs must account for the diversity of responses elicited by these compounds.

The biologic and toxic potencies of PCDDs and PCDFs are dependent on the animal species used, the response being investigated, and the structure of the administered compound. The interspecies sensitivity to the acute lethal effect of 2,3,7,8-TCDD is dramatic and is evidenced by the following $LD_{50}$ values: guinea pig (0.6–2.0 μg/kg), rat (22–45 μg/kg), chicken (25–50 μg/kg), monkey (70 μg/kg), rabbit (115 μg/kg), dog (100–200 μg/kg), mouse (114–284 μg/kg), bull frog ( > 1000 μg/kg), and hamster (1157–5000 μg/kg) (Kociba and Schwetz, 1982a,b). These results indicate a greater than 5000-fold difference in the $LD_{50}$ between the guinea pig and the hamster, but this marked species variability, as measured by the acute $LD_{50}$ values, may not be typical of other biological responses. For example, rats and hamsters are equally sensitive to the induction of hepatic microsomal thyroxine glucuronyl transferases by 2,3,7,8-TCDD (Henry and Gasiewicz, 1987). Gasiewicz and coworkers (1986) have reported that the $ED_{50}$ values for thymic atrophy and hepatic microsomal ethoxycoumarin *O*-deethylase (ECOD) induction in hamsters were 100 μg/kg and 0.5 μg/kg, respectively. These values are comparable to data for 2,3,7,8-TCDD-elicited thymic atrophy in the rat (Safe, 1987) and enzyme induction data in the rat and mouse (Poland and Glover, 1974; Greenlee and Poland, 1978; Safe, 1987). The implication of these results is that there may not be a wide species variation in susceptibility to some of the toxicities elicited by 2,3,7,8-TCDD and related compounds.

Wasting syndrome and lymphoid involution are the two most common toxic responses to 2,3,7,8-TCDD and related compounds which are observed in most animal species. In contrast, many other responses caused by halogenated aryl hydrocarbons are highly species-specific (Table 2). Although the mechanism(s) responsible for the species-specific effects of PCDDs and PCDFs are unknown, the observed mouse strain differences suggest that genetic factors strongly influence some of the biological responses. The issue of variability among species contributes to the uncertainty involved in assessing the risk that PCDDs and PCDFs pose for humans.

### *1.2 Possible Initial Mechanisms of Action*

#### Covalent Binding to DNA

In general, PCDDs and PCDFs elicit the same species-specific biological responses, although the compounds differ markedly in their potency. Therefore, it has been assumed that the compounds share a common mechanism of action. The vast majority of studies have employed 2,3,7,8-TCDD (the most potent

**Table 2.** Toxicity of 2,3,7,8-TCDD and related compounds: Species differences[a]

| Effect | Species | | | | | |
|---|---|---|---|---|---|---|
| | Human | Monkey | Guinea Pig | Mouse | Chicken | Rat |
| Body weight loss | | + | + | + | + | + |
| Chloracne and related dermal lesions | + | + | ND | + [b] | ND | ND |
| Developmental toxicity | | + | | + | + | + |
| Edema | | + | ND | + | + | ND |
| Liver damage | | + | ND | + | + | + |
| Teratogenicity | | ND | ND | + | + | ND |
| Thymic atrophy | | + | + | + | + | + |

[a] Reviewed in Poland and Knutson (1982); [b] observed only in hairless HRS/J mice; + = yes; ND = not demonstrated

compound) as a prototype for mechanistic studies. This compound is metabolized slowly (Neal et al., 1984) and several of its hydroxylated metabolites are much less toxic than the parent compound (Mason and Safe, 1986). Furthermore, 2,3,7,8-TCDD fails to undergo appreciable metabolic activation to a form that can bind covalently to nucleic acids or proteins (Poland and Glover, 1979). These findings imply that the unmetabolized parent compound elicits the biological responses and that the mechanism, at least for 2,3,7,8-TCDD, is unlikely to involve direct damage to DNA resulting from covalent binding.

## Direct Damage to Cellular DNA (Mutagenesis)

The literature regarding the genetic toxicity of 2,3,7,8-TCDD and related compounds has been reviewed recently (Kociba, 1984; U.S. Environmental Protection Agency, 1985; Giri, 1986; Shu et al., 1987). In general, the halogenated aromatic hydrocarbons have shown little ability to induce gene mutations in test systems using bacteria, yeast, mammalian cells in culture, or Drosophila. The PCDDs and PCDFs, in particular, are inactive in short-term microbial assays for mutagenesis (Wassom et al., 1978; Geiger and Neal, 1981). While some studies have suggested the possibility that 2,3,7,8-TCDD may induce cytogenetic damage (Jackson, 1972; Green et al., 1977; Loprieno et al., 1982), other studies (Meyne et al., 1985) have not. Most in vitro genetic toxicity studies suggest that 2,3,7,8-TCDD does not act via direct damage to cellular DNA (Shu et al., 1987). In the studies by Jackson (1972), 2,3,7,8-TCDD was reported to produce chromosomal abnormalities in African lily plants. In non-human mammalian systems, Loprieno et al. (1982) reported cytogenetic changes in the bone marrow of mice given a single dose of 10 µg/kg 2,3,7,8-TCDD by intraperitoneal injection, but no changes were observed in the bone marrow of rats similarly treated at lower doses. On the other hand, Green et al. (1977) reported a weak response in chromosomal aberrations in the bone marrow of rats treated orally with 2,3,7,8-TCDD, but the authors included no controls in their study. Meyne et al. (1985) reported no cytogenetic changes in bone marrow of mice administered 50–150 µg/kg of 2,3,7,8-TCDD.

Several cytogenetic studies have been conducted in humans alleged to have had occupational or environmental exposures to 2,3,7,8-TCDD. The preponderance of evidence from these studies indicates that exposures have not been associated with cytogenetic changes in humans (Shu et al., 1987).

While scattered positive results have been reported in some assays in vitro, the weight of evidence based on in vivo studies suggests that 2,3,7,8-TCDD is not genotoxic in mammalian systems.

## Receptor Mediated Activity

Two lines of evidence implicate an intracellular receptor protein in the mechanism of 2,3,7,8-TCDD action. First, studies of inbred mouse strains revealed a difference in their sensitivities to 2,3,7,8-TCDD, as measured by the induction of hepatic aryl hydrocarbon hydroxylase (AHH) activity (Poland and Glover, 1973, 1974, 1975; Poland et al., 1974). This difference correlated with the presence in liver tissue of an intracellular protein that bound [$^{3}$H]-2,3,7,8-TCDD saturably and with high affinity, and thus possessed the properties of a 2,3,7,8-TCDD "receptor" (Poland et al., 1976). Second, studies with individual PCDD and PCDF congeners revealed that their affinities for the receptor paralleled their potencies as AHH inducers, suggesting that the receptor participated in the process of enzyme induction (Poland and Knutson, 1982; Safe, 1986). The 2,3,7,8-TCDD receptor is known as the Ah receptor because it also mediates enzyme induction in response to certain aromatic hydrocarbons such as 3-methylcholanthrene (3-MC) and benz(a)anthracene (Thomas et al., 1972). The genetic locus that determines responsiveness to aromatic hydrocarbons is designated the *Ah* locus and presumably encodes the 2,3,7,8-TCDD receptor protein. Toxic effects of 2,3,7,8-TCDD which also segregate with the *Ah* locus in inbred mice, include cleft palate (Poland and Glover, 1980), thymic atrophy (Poland and Glover, 1980), suppression of cell-mediated and humoral immunity (Vecchi et al., 1980, 1983; Nagarkatti et al., 1984), myelotoxicity (Luster et al., 1985; Hong et al., 1987), hepatic porphyria (Jones and Sweeney, 1980; Greig et al., 1984), epidermal hyperplasia and hyperkeratosis (Knutson and Poland, 1982), and lethality (Gasiewicz et al., 1983; Chapman and Schiller, 1985) and, therefore, presumably represent responses mediated by the 2,3,7,8-TCDD receptor. Although previous observations suggested that the Ah receptor mediated the tumor-promoting action of 2,3,7,8-TCDD in mice, Poland and coworkers (1982) demonstrated that the receptor alone was not sufficient to evoke the response since the promoting effect occurred only in animals with *hr/hr* alleles in their genetic makeup.

Genetic evidence for the *Ah* locus exists primarily in mice; however, other species, including humans, contain a 2,3,7,8-TCDD-binding protein with properties similar to the mouse receptor (see Sect. 2). In addition, there is a good correlation between the structure-activity relationships for binding to a receptor protein and toxicity in other experimental animals (see Sect. 3). Therefore, these other species presumably possess a functional equivalent of the mouse *Ah* locus.

### *1.3 Receptor Model of Initial Toxic Mechanism*

The three models considered above are the more biologically plausible mechanisms of initial interaction between PCDDs/PCDFs and cellular constituents. Investigations of all three have led to the exclusion of covalent binding to DNA, direct chromosomal damage, or DNA defect as the most likely first stage of toxic effect. The remainder of this chapter reviews in detail the experimental evidence which is consistent with the mechanistic model for the action of 2,3,7,8-TCDD as shown in Fig. 1.

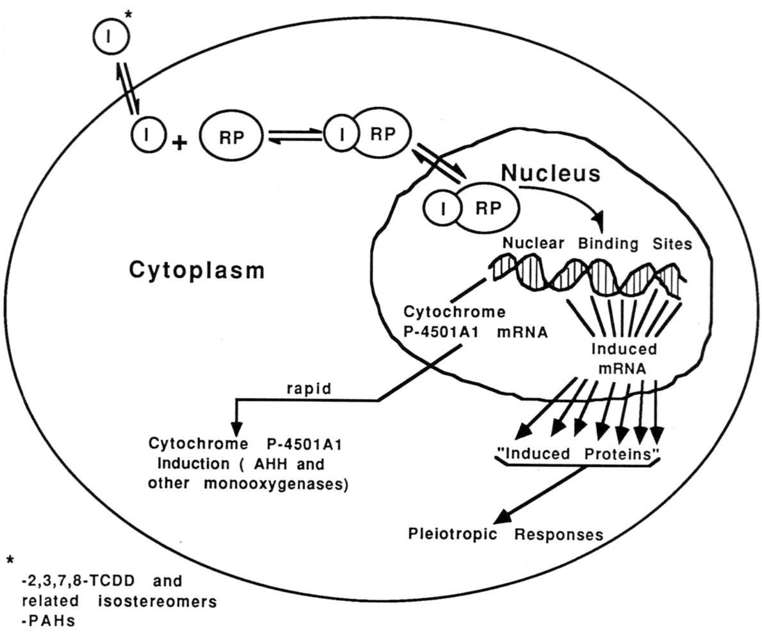


**Fig. 1.** Proposed mechanism of action of 2,3,7,8-TCDD and related compounds

## 2 Molecular Properties and Binding Characteristics of the Ah Receptor

### *2.1 Molecular Properties of Ah Receptor-Ligand Complexes*

If we presume that the Ah receptor mediates the biological responses to PCDDs, PCDFs, and related aryl hydrocarbons, then it is important to examine the properties and actions of this protein to ascertain if and how they may determine species susceptibility. Several groups have investigated the molecular properties of the Ah receptor from a variety of animal species and tissues. For comparative purposes, the most useful physicochemical data have been obtained using [$^3$H]-2,3,7,8-TCDD as the radioligand. Table 3 summarizes these data (Safe, 1988).

Although certain species-specific differences appear to exist (see also Sect. 2.2), some common features have been observed. As defined by the frictional ratio, the protein appears to be asymmetric. In solutions of low salt concentrations ($\leqslant$ 0.1 M KCl), it is present as a high molecular weight complex of 257,000–277,000 daltons (sedimentation coefficients of 7–9), whereas in high salt solutions (0.4 M KCl) the molecular weight ranges from 100,000–120,000 dal-

**Table 3.** Molecular properties of the Ah receptor

| Source (Ref.)* | Sedimentation Coeff. ($S_{20,w}$) | Stokes Radius (nm) | Relative Mass, Daltons | Frictional Ratio (f/fo) | Axial Ratio (a/b) |
|---|---|---|---|---|---|
| Rat hepatic cytosol | | | | | |
| 0.1 M KCl (1) | 8.8 ± 0.05 | 7.0 ± 0.21 | 257,000 ± 7,700 | 1.7 ± 0.03 | 12.4 ± 0.69 |
| 0.4 M KCl (1)[a] | 5.6 ± 0.58 | 5.2 ± 0.24 | 121,000 ± 5,000 | 1.6 ± 0.05 | 11.3 ± 1.00 |
| 0.4 M KCl (2, 3)[b] | ~ 4–5 | ~ 6.0 | ~ 100,000 | 1.7–1.8 | 12–15 |
| C57BL/6 mouse hepatic cytosol | | | | | |
| 0.1 M KCl (1) | 9.4 ± 0.57 | 7.1 ± 0.12 | 277,000 ± 4,800 | 1.7 ± 0.02 | 12.2 ± 0.04 |
| 0.4 M KCl (1) | 9.7 ± 0.20 | 6.8 ± 0.15 | 274,000 ± 5,300 | 1.6 ± 0.02 | 11.0 ± 0.40 |
| | 5.5 ± 0.02 | 5.2 ± 0.19 | 105,000 ± 3,800 | 1.7 ± 0.04 | 11.9 ± 0.85 |
| 0.4 M KCl (4) | 4–5 | 6.1–6.2 | 100,000–120,000 | 1.7–1.8 | 12–14 |
| Mouse Hepa 1c1c7 cells (4)[b] | 7.5–8.0 | 8.2–8.7 | 260,000–300,000 | 1.6–1.7 | 12–13 |
| Human placenta (5)[b] | 8.5 ± 1.0[c] | | | | |

* (1) Denison et al., 1986a; (2) Wilhelmsson et al., 1986; (3) Poellinger et al., 1983; (4) Cuthill et al., 1987; (5) Manchester et al., 1987
[a] Molybdate partially stabilizes high molecular weight complex (Denison et al., 1986b; Wilhelmsson et al., 1986); [b] unaffected by molybdate;
[c] 0.1 and 0.4 M KCl in the presence of molybdate

tons (sedimentation coefficients of 4–6). It is not known whether the higher molecular weight complex is a multimer of the same protein, or whether it exists in combination with other proteins. However, the high molecular weight ligand-receptor complexes isolated from C57BL/6J mice and mouse Hepa 1c1c7 cells are highly stable even in high concentration salt solutions (Denison et al., 1986a; Cuthill et al., 1987), and, furthermore, molybdate has been shown to stabilize the high molecular weight complexes isolated from rat liver and human placenta (Denison et al., 1986b; Wilhelmsson et al., 1986; Manchester et al., 1987). The significance of the latter observation is that many steroid hormone receptors are affected in a similar manner by molybdate. Comparable physicochemical properties have been observed using other radioligands. Some of these ligands include 2,3,7,8-TCDF, 1,2,3,7,8-PeCDF, 1,2,3,4,7,8-1,2,3,6,7,8-HxCDF, 3-MC, dibenzo[a,h]anthracene, and benzo[a]pyrene (Hannah et al., 1981; Okey and Vella, 1982; Poellinger et al., 1983; Okey et al., 1984a,b; Farrell et al., 1987; Manchester et al., 1987).

The hydrophobic, polyanionic, and DNA binding properties of the Ah receptor-2,3,7,8-TCDD complexes have also been reported and compared to those of the steroid hormone receptors, in particular the glucocorticoid receptor (Carlstedt-Duke et al., 1981; Poellinger and Gullberg, 1985; Hannah et al., 1986; Wilhemsson et al., 1986; Gasiewicz and Bauman, 1987). In general, the properties of the two types of receptor proteins are similar, although the Ah receptor appears to be more hydrophobic (Poellinger and Gullberg, 1985). In addition, both proteins undergo ligand and temperature-dependent changes in conformation and overall charge that appear to be necessary for the binding to DNA-cellulose/Sepharose$^{R}$ (Hannah et al., 1986; Wilhelmsson et al., 1986; Gasiewicz and Bauman, 1987). It was also reported that mild proteolysis of the receptor with trypsin or chymotrypsin did not reduce the ligand-binding properties of the receptor but significantly reduced the interaction of the bound complex with DNA (Carlstedt-Duke et al., 1981). These data are consistent with the presence of at least two domains on the Ah receptor protein: a ligand-binding domain and a DNA-binding domain. These results are also in agreement with an earlier study noting a temperature-dependent nuclear association of the occupied receptor (Okey et al., 1980). Furthermore, there is a receptor-defective variant of mouse hepatoma cells in which relatively few 2,3,7,8-TCDD-receptor complexes form, but these interact normally with the nucleus. In another class of variants, although the formation of the complex appears normal, the ligand-receptor complexes fail to interact normally with the nucleus, and the cells do not respond to 2,3,7,8-TCDD (Hankinson, 1979; Miller and Whitlock, 1981; Legraverend et al., 1982; Miller et al., 1983). Although the actual biochemical and/or conformational changes that take place in the receptor to convert it to a chromatin- binding species are unknown, recent evidence suggests the simultaneous presence of multiple forms of the Ah receptor whose interconversion may be mediated by an ATP-dependent mechanism (Gasiewicz and Bauman, 1987; Hannah, 1988). The maintenance of reduced sulfhydryl groups has also been shown to be important for ligand binding to the Ah receptor (Denison et al.,

1987; Kester and Gasiewicz, 1987; Henry et al., 1988) and the interaction of the ligand-receptor complex with DNA-Sepharose[R] (Henry et al., 1988).

Although biochemically and functionally the properties of the Ah receptor protein are similar to those of certain steroid hormone receptors, in particular the glucocorticoid receptor, the data presented to date indicate that the proteins are not identical. Binding specificities appear to be mutually exclusive (Poland et al., 1976; Neal et al., 1979), and a monoclonal antibody directed against the glucocorticoid receptor does not react with the Ah receptor (Poellinger et al., 1983). The similarity of the properties and functions of this protein to those of the steroid hormone receptors implies that the Ah receptor is a gene regulatory protein having some normal function in basic cellular processes. This further implies that the PCDDs and PCDFs interfere with the normal endogenous ligand and/or provide, through a series of molecular events in which the properties of the receptor are changed, an inappropriate signal for the alteration of gene expression. However, the endogenous ligand, if any, for the Ah receptor, and its role in the regulatory processes of the cell, remain unestablished.

## 2.2 *Ah Receptor: Radioligand Binding Affinities, Species Differences, and Distribution*

Using radiolabeled 2,3,7,8-TCDD, the Ah receptor was first identified in hepatic cytosol from C57BL/6 mice (Poland et al., 1976). The receptor-ligand binding interaction was saturable (approximately 84 fmol/mg cytosolic protein) and the apparent equilibrium dissociation constant, $K_d$, as determined by Scatchard analysis, was 0.27 nM. Subsequent studies by a number of laboratories have identified the Ah receptor in several animal species and in cultured mammalian cells (Okey et al., 1979, 1980; Gasiewicz and Rucci, 1984a; Whitlock and Galeazzi, 1984; Roberts et al., 1985; Denison et al., 1986c; Cuthill et al., 1987). Table 4 summarizes the estimated receptor levels and $K_d$ values for hepatic

**Table 4.** [$^3$H]-2,3,7,8-TCDD Ah receptor binding capacity and dissociation constant values in several species and cells in culture (Gasiewicz and Rucci, 1984a; Denison et al., 1986c)

| Species | Total Specific Binding, femtomoles/mg cytosolic protein | $K_d$, nanomoles/liter |
|---|---|---|
| Chick Embryo (7DI)[a] | 70 ± 16[b] | 0.20 ± 0.06 |
| Gerbil | 87 ± 8 | 0.20 ± 0.04 |
| Guinea pig | 114 ± 1, 59 ± 11 | 0.16 ± 0.02, 0.06 ± 0.02 |
| Hamster | 67 ± 22, 78 ± 10 | 0.33 ± 0.07, 0.22 ± 0.02 |
| Monkey | 42 ± 8 | 0.26 ± 0.04 |
| Mouse (C57BL/6) | 92 ± 10, 74 ± 10 | 0.52 ± 0.09, 0.29 ± 0.04 |
| Mouse ($B6D2F_1$/J) | 23 ± 2 | |
| Rat | 97 ± 5, 61 ± 23 | 0.22 ± 0.05, 0.12 ± 0.03 |

[a] 7DI = seven day incubation; [b] mean ± standard deviation

cytosol from different animals. The results from these laboratories illustrate the relatively narrow range of $K_d$ values for the receptor-ligand complexes and the broader range of hepatic receptor levels. There is no obvious relationship between species susceptibility to the biologic effects of 2,3,7,8-TCDD and the amount of Ah receptor contained in hepatic tissue.

Poland and coworkers (1987) recently reported differences in Ah receptor binding affinities in genetically inbred mice. Analysis of specific binding of [$^3$H]-2,3,7,8-TCDD to the Ah receptor from two classes of responsive mice (designated $Ah^{b-1}$ and $Ah^{b-2}$) showed that the $K_d$ values and receptor levels in mice carrying the $Ah^{b-1}$ allele were 0.4–0.7 nM and 130–160 fmol/mg cytosolic protein, respectively. In contrast, the $K_d$ values and receptor levels for strains of mice carrying the $Ah^{b-2}$ allele were 0.8–1.3 nM and 80–110 fmol/mg cytosolic protein, respectively. These studies are of interest since certain mouse strains possessing a receptor with high affinity for 2,3,7,8-TCDD are more susceptible to 2,3,7,8-TCDD-elicited toxicity (Poland and Glover, 1980). Although the form of the receptor found in human placenta apparently has a relatively low affinity ($K_d$ = 5–8 nM) for [$^3$H]-2,3,7,8-TCDD (Manchester et al., 1987), no firm conclusion can yet be drawn as to the potential susceptibility of this or other human tissue to this compound.

In addition to [$^3$H]-2,3,7,8-TCDD, several other radiolabeled ligands have been used to examine receptor numbers and relative binding affinites (Hannah et al., 1981; Okey and Vella, 1982, 1984; Okey et al., 1983, 1984a,b; Manchester et al., 1987; Farrell et al., 1987). In general, the relative affinity of various halogenated dibenzo-*p*-dioxins and dibenzofurans for the Ah receptor, as determined by competitive binding with [$^3$H]-2,3,7,8-TCDD, correlates with their toxic potency in experimental animals. However, certain PCDD and PCDF isomers have been found to be less toxic than would otherwise be predicted from the binding affinity data. This may be due to increased rates of metabolism of the ligand in vivo and/or antagonistic properties (see Sect. 3 for detailed discussion).

Poland and Glover (1987) applied a photoaffinity labeling technique to examine the molecular weights of the Ah receptor from several strains of mice and other vertebrate species. The apparent molecular weights of the photolabeled proteins ranged from 101 to 124 kDa. Although these results confirm that there are differences in the structure of the Ah receptor from different species, the relationship, if any, to binding affinity and/or susceptibility to toxicity is not clear.

Both linear and non-linear Scatchard plots for the binding of [$^3$H]-2,3,7,8-TCDD to the Ah receptor have been reported (Poland et al., 1976; Okey et al., 1979; Poellinger et al., 1983; Gasiewicz and Rucci, 1984a; Farrell and Safe, 1987). If positive cooperativity were present (as might be suggested by a non-linear Scatchard plot), it would implicate the occurrence of a ligand-induced conformational change in the receptor to a higher affinity state characterized by a slower rate of dissociation. In turn, the slower rate of dissociation may prolong the biological action of 2,3,7,8-TCDD. Hill plot analysis of several binding isotherms gave slopes of 1.0 which supported the lack of positive cooperativity in this binding process (Gasiewicz and Rucci, 1984a; Farrell and Safe, 1987). However, extremely slow rates of dissociation of [$^3$H]-2,3,7,8-TCDD from the rat hepatic Ah receptor have been observed (Farrell and Safe, 1987; Henry et al., 1988). Although the data are equivocal regarding the presence of positive cooperativity in the binding process, the slow dissociation of 2,3,7,8-TCDD from the Ah receptor may be responsible in part for the sustained alteration of gene expression (Poland and Knutson, 1982) and may be an important factor in the toxicity of this compound.

Initial studies reported that the unoccupied (i.e., unoccupied by 2,3,7,8-TCDD) Ah receptor was located primarily, if not exclusively, in the soluble (cytosolic) fraction following tissue homogenization (Poland et al., 1976). However, using volume dilution techniques, it was observed that unoccupied receptor was primarily nuclear in origin (Whitlock and Galeazzi, 1984). Denison and coworkers (1986e) confirmed these results, but noted that the distribution of the Ah receptor was identical to that of several cytosolic marker enzymes. In addition, enucleation of cells by cytochalasin B treatment indicated the presence of the unoccupied receptor only in the soluble fraction (Gudas et al., 1986). These latter studies suggest that the unoccupied receptor is primarily a cytosolic protein. However, these data do not exclude the possibility of a nuclear-cytosolic equilibrium process. Although the details of the distribution of the unoccupied form of the receptor do not necessarily lead to a greater understanding of the mechanism of action of PCDDs and PCDFs, this information coupled with other data emphasizes the multistep molecular processes that ultimately lead to the observed biological effects. For example, it is clear that following treatment of cells or experimental animals with [$^3$H]-2,3,7,8-TCDD, the major percentage of the ligand-receptor complex becomes associated with the nuclear fraction (Okey et al., 1979, 1980; Gasiewicz and Rucci, 1984b). Thus 2,3,7,8-TCDD binding to the receptor in vivo induces a biochemical/conformational change to result in a protein with higher affinity for nuclear chromatin.

The tissue distribution of the Ah receptor in several species has also been reported (Carlstedt-Duke, 1979; Mason and Okey, 1982; Okey et al., 1983; Gasiewicz and Rucci, 1984a); Table 5 summarizes results from one of these studies. Although there were major interspecies differences in receptor levels in various tissues and organs, the liver, lung, and thymus appeared to contain the highest levels of protein. It should be pointed out that the ability to detect [$^3$H]-2,3,7,8-TCDD binding to the Ah receptor may be a function of the assay used. For example, Manchester and coworkers (1987) noted that binding to a lower affinity form of the receptor contained in human placenta was low or undetectable by some of the conventional adsorption assays previously used for other tissues and species, but was detectable by sucrose density gradient analysis. Thus, previous inability to detect the presence of the Ah receptor may be due, in part, to the predominant presence of lower affinity forms.

From the data that are available there does not appear to be a clear relationship between tissue- and species-specific differences in Ah receptor levels

**Table 5.** Tissue concentrations of receptors for TCDD[a] (Gasiewicz and Rucci, 1984a)

| | Guinea Pig | Hamster | Rat | Mouse |
|---|---|---|---|---|
| Liver | $59 \pm 11$ (8)[b] | $67 \pm 22$ (7) | $61 \pm 23$ (7) | $74 \pm 10$ (4) |
| Thymus | $47 \pm 7$ (8) | $5 \pm 6$ (4) | $121 \pm 30$ (4) | $24 \pm 2$ (3) |
| Lung | $86 \pm 28$ (4) | $35 \pm 20$ (4) | 76 (2) | — |
| Kidney | $24 \pm 19$ (4) | 13 (2) | — | — |
| Spleen | 17 (2) | $6 \pm 5$ (3) | — | — |
| Adrenal | $3 \pm 3$ (3) | N.D.[c] (2) | — | — |
| Duodenum | 18 (2) | N.D. (2) | — | — |
| Heart | 16 (2) | N.D. (2) | — | — |
| Testes | $50 \pm 7$ (4) | N.D. (2) | — | — |
| Muscle | N.D. (2) | N.D. (2) | — | — |
| Pancreas | N.D. (2) | N.D. (2) | — | — |

[a] Values are expressed in femtomoles per mg of cytosolic protein; [b] mean ± standard deviation of (n) determinations; [c] N.D. = not detectable

and the relative susceptibility to toxicity. Whether or how the presence of lower affinity forms of the receptor affects susceptibility in all tissues and species is also not known. However, a relationship between the presence of a high affinity form of the receptor in certain inbred strains of mice and their increased susceptibility to 2,3,7,8-TCDD-elicited enzyme induction and toxicity has been established (Poland and Glover, 1980).

## *2.3 Stereoselective Ah Receptor-Ligand Interactions*

The structure-activity relationships (SAR) observed for the receptor binding affinities of PCDDs and PCDFs further support the role of the Ah receptor in mediating the responses elicited by these compounds. Poland and colleagues (1976) first reported the stereoselective interactions of halogenated aryl hydrocarbons with the Ah receptor using a competitive binding assay and [$^3$H]-2,3,7,8-TCDD as the displaceable radioligand. The effects of structure on the competitive receptor binding affinities of diverse halogenated dibenzo-*p*-dioxins and dibenzofurans using rat hepatic cytosol and [$^3$H]-2,3,7,8-TCDD as the radioligand have also been reported (Poland and Knutson, 1982; Mason et al., 1985, 1986, 1987; Safe, 1986, 1987). For the PCDDs, the compounds with the highest binding affinities were substituted in at least 3 of the 4 lateral 2, 3, 7, and 8 positions and removal of one or more of the lateral substituents or the addition of a non-lateral halogen group gave congeners with decreased competitive receptor binding affinities.

The dibenzofuran ring system possesses only a single axis of symmetry and there are 4 different positions on each ring, namely C-1 (or C-9), C-2 (or C-8), C-3 (or C-7), and C-4 (or C-6). Extensive SARs for PCDFs as competitive ligands for the rat hepatic receptor have been reported (Bandiera et al., 1984;

Mason et al., 1985; Safe, 1986) and these data demonstrate that chlorine substitution at all 4 positions in the dibenzofuran ring contributes differentially to the binding affinities of the resultant congeners. These studies show that the individual contribution of chlorine substitution to the biologic and toxic potencies of PCDFs follow the order C-3 (or C-7) > C-2 (or C-8) > C-4 (or C-6) > C-1 (or C-9).

Figure 2 summarizes the structures of several substituted halogenated aromatics which have been utilized to probe the physicochemical factors which govern the ligand-Ah receptor binding site interactions and the heterologous nature of the Ah receptor between species (Bandiera et al., 1983; Denomme et al., 1985, 1986a; Romkes et al., 1987b). Multiple parameter linear regression analysis of the hepatic receptor binding data for the 2-substituted-3,7,8-trichlorodibenzo-*p*-dioxins gave Eqs. 1–4:

$$pEC_{50}\ (\text{rat}) = 7.19 + 0.600\pi + 0.255\ \Delta Es - 1.683\ HB \quad (1)$$

$$pEC_{50}\ (\text{mouse}) = 6.365 + 1.641\pi + 1.206\ \sigma^0 \quad (2)$$

$$pEC_{50}\ (\text{hamster}) = 7.416 + 1.026\pi + 0.509\ \Delta Es + 0.748\ \sigma^0 \quad (3)$$

$$pEC_{50}\ (\text{guinea pig}) = 6.892 + 1.035\pi \quad (4)$$

where $\pi$, $\Delta Es$, HB, and $\sigma^0$ represent substituent lipophilicity, Taft constants, hydrogen binding capacity, and electronegativity, respectively. The differences in Eqs. 1–4 confirm the heterologous nature of the Ah receptor between mammalian species. The receptor binding data analysis for all the halogenated aromatic ligands (Fig. 2) showed the importance of substituent $\pi$ values for facilitating the interactions of these highly lipophilic compounds with the receptor. However, the attractive forces associated with $\pi$, $\Delta Es$, HB, and $\sigma^0$ do not explain the tight binding and slow rate of dissociation observed for the 2,3,7,8-TCDD-Ah receptor complexes (Farrell and Safe, 1987; Henry et al., 1988). This suggests that other factors such as receptor protein conformational changes play an important part in the binding of halogenated aromatic hydrocarbons with the Ah receptor.

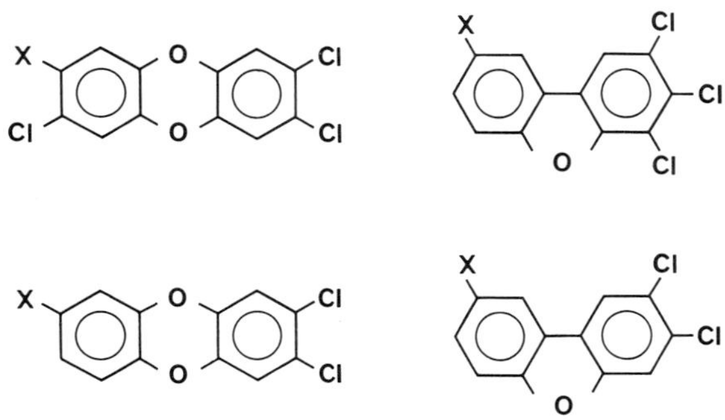


**Fig. 2.** Substituted halogenated aryl hydrocarbons used for quantitative structure activity relationship (QSAR) studies

Several additional studies have reported the Ah receptor interactions of structurally diverse organic compounds, including 5,6-benzoflavone and substituted indoles (Gillner et al., 1985), several polynuclear aromatic hydrocarbons (Piskorska-Pliszczynska et al., 1986), substituted diaryltriazenes (Sweatlock and Gasiewicz, 1986), and selected Sudan dyes (Lubet et al., 1983). Analysis of the three dimensional structures of several of these compounds indicated that the non-halogenated aryl hydrocarbons which exhibited relatively high Ah receptor binding could be fitted into a rectangle with dimensions of $6.8 \times 13.7$ Å (Gillner et al., 1985). It has also been suggested that in addition to steric and size requirements, substituent polarizability may also be an important structural determinant for facilitating ligand-Ah receptor interactions (Long et al., 1987). Inasmuch as the sustained alteration of gene expression by 2,3,7,8-TCDD and related compounds may be the result of long term receptor occupancy (Poland and Knutson, 1982), a more comprehensive quantitative study of other ligand-receptor interactions may help to further delineate the molecular forces which govern these interactions.

## 2.4 *Ah Receptor: Ontogeny, Regulation, and Relationships Between Receptor Levels and Biological Response*

The ontogeny of the receptor in hepatic and extrahepatic tissues has also been investigated (Carlstedt-Duke et al., 1979; Kahl et al., 1980; Gasiewicz et al., 1984; Densison et al., 1986d). In Sprague-Dawley rats, hepatic receptor levels were relatively low prior to birth ( $<$ 50 fmol/mg cytosolic protein); however, these levels increased to greater than 250 fmol/mg in 8 day-old animals and were maintained through day 21 (Kahl et al., 1980; Gasiewicz et al., 1984). In 28 to 70 day-old animals, the values decreased to approximately 100 fmol/mg cytosolic protein. A similar pattern of age-dependency in hepatic Ah receptor levels was found in other species examined (Kahl et al., 1980). A comparable age-dependent variation in lung receptor levels was also observed, whereas thymic receptor levels were relatively constant (Gasiewicz et al., 1984). The ontogeny of the Ah receptor in chickens was markedly different than in rats. Maximum receptor levels were observed in 5 to 9 day-old chick embryos ( $>$ 40 fmol/mg cytosolic protein) whereas in older embryos and in neonates the receptor levels were generally lower ( $<$ 20 fmol/mg) (Denison et al., 1986c). The biological significance of age-dependent Ah receptor levels is unclear. Although there appeared to be a general correspondence between the level of Ah receptor and enzyme inducibility in the rat, mouse, and rabbit (Kahl et al., 1980), the AHH inducibility and toxic susceptibility of chickens to 2,3,7,8-TCDD is high regardless of receptor level (Denison et al., 1986d; Sawyer et al., 1986).

Recent studies have demonstrated that hepatic receptor levels can be significantly modulated by a variety of xenobiotics including Aroclor 1254, 2,2′,4,4′,5,5′-hexachlorobiphenyl, and several other polychlorinated biphenyls

(Denomme et al., 1986b; Bannister and Safe, 1987), phenobarbital (Okey and Vella, 1984), 1,4-bis(2-(3,5-pyridyloxy))benzene (Lesca et al., 1987), and *trans*-4-acetylaminostilbene (Gottlicher and Cikryt, 1987). Although there are no apparent common structural features for these compounds, their apparent $K_d$ values are not significantly different and all have been shown to increase rodent hepatic receptor levels approximately 100–200%. A similar observation has been made in rats following treatment with 2,3,7,8-TCDD (Sloop and Lucier, 1987), but no alteration in hepatic receptor levels was observed in 2,3,7,8-TCDD-treated hamsters (Gasiewicz and Rucci, 1987). It has been reported that male Wistar rats and C57BL/6J mice treated with 2,2′,4,4′,5,5′-hexachlorobiphenyl are more responsive to the induction of AHH and ethoxyresorufin *O*-deethylase (EROD) activities by 2,3,7,8-TCDD; however, there appeared to be no toxicological significance to the increased receptor concentrations (Bannister and Safe, 1987; Leece et al., 1987). Hormone-mediated modulation (up- and down-regulation) of steroid hormone receptor levels has frequently been reported and the 2,3,7,8-TCDD-mediated increase in Ah receptor concentrations is not entirely unexpected. However, the effectiveness of phenobarbital (PB) and several PCB congeners that exhibit PB-type monooxygenase induction activity is surprising because the overlap in the biologic and toxic activities of these xenobiotics and 2,3,7,8-TCDD is minimal (Safe, 1984).

Although a number of investigators have observed alterations in the tissue concentrations of the Ah receptor with age and following various chemical treatments, the biological significance of these alterations with respect to susceptibility to the toxic effects of the PCDDs and PCDFs is not clear.

## 3 Mechanism of Action of 2,3,7,8-TCDD and Related Compounds: Role of the Ah Receptor

### *3.1 PCDDs and PCDFs: Structure-Induction Relationships*

The quantitative SARs for PCDFs (Table 6a) and PCDDs (Table 6b) as inducers of AHH and EROD in rat hepatoma H-4-II E cells and in guinea pig and rat hepatic microsomes have been reported (Bandiera et al., 1984; Mason et al., 1985, 1986; Holcomb et al., 1988). The in vivo and in vitro SARs for enzyme induction and toxicity (see Sect. 3.2) and for receptor binding were comparable, and these correlations provide further support for the role of the Ah receptor in these processes. However, it was apparent that there was not a linear correlation between the $-\log EC_{50}$ (in vitro) or $\log ED_{50}$ (in vivo) values for enzyme induction or toxicity and the $-\log EC_{50}$ values for receptor binding. The greatest differences between competitive receptor binding affinities and monooxygenase induction potencies were observed with 1,3,6,8- and 2,4,6,8-TCDF (Keys et al., 1986). These compounds exhibited relatively high receptor binding

**Table 6a.** Structure activity relationships of polychlorinated dibenzofurans as inducers of aryl hydrocarbon hydroxylase (AHH) and ethoxyresorufin *O*-deethylase (EROD) in rat hepatoma H-4-II E cells in culture

| Congeners | AHH $EC_{50}$, mol/liter | EROD $EC_{50}$, mol/liter |
|---|---|---|
| Dibenzofuran | ND[a] | ND |
| 2-[b] | ND | ND |
| 3- | ND | ND |
| 4- | $1.0 \times 10^{-5}$ | $1.7 \times 10^{-5}$ |
| 2,6- | $6.2 \times 10^{-5}$ | $6.3 \times 10^{-5}$ |
| 2,8- | $4.0 \times 10^{-5}$ | $4.0 \times 10^{-5}$ |
| 2,3- | $2.2 \times 10^{-6}$ | $4.8 \times 10^{-6}$ |
| 1,3,8- | $1.9 \times 10^{-5}$ | $3.0 \times 10^{-5}$ |
| 2,6,7- | $2.8 \times 10^{-6}$ | $3.1 \times 10^{-6}$ |
| 1,3,6- | $2.5 \times 10^{-6}$ | $3.4 \times 10^{-6}$ |
| 2,3,8- | $2.5 \times 10^{-6}$ | $1.6 \times 10^{-6}$ |
| 2,3,4- | $1.5 \times 10^{-7}$ | $2.5 \times 10^{-7}$ |
| 1,2,3,6- | $> 10^{-4}$ | $> 10^{-4}$ |
| 1,2,3,7- | $2.7 \times 10^{-5}$ | $6.3 \times 10^{-5}$ |
| 1,2,4,8- | $1.2 \times 10^{-5}$ | $9.3 \times 10^{-5}$ |
| 2,3,4,6- | $1.3 \times 10^{-6}$ | $1.1 \times 10^{-6}$ |
| 2,3,6,8- | $1.0 \times 10^{-6}$ | $7.8 \times 10^{-7}$ |
| 2,3,4,8- | $4.1 \times 10^{-8}$ | $3.8 \times 10^{-8}$ |
| 2,3,7,8- | $3.9 \times 10^{-9}$ | $2.0 \times 10^{-9}$ |
| 1,2,4,6,7- | $3.2 \times 10^{-7}$ | $3.5 \times 10^{-7}$ |
| 1,2,3,4,8- | $2.1 \times 10^{-7}$ | $1.6 \times 10^{-7}$ |
| 1,2,4,7,8- | $1.1 \times 10^{-7}$ | $1.5 \times 10^{-7}$ |
| 1,2,3,7,9- | $8.6 \times 10^{-8}$ | $8.6 \times 10^{-8}$ |
| 1,2,4,7,9- | $3.8 \times 10^{-8}$ | $3.8 \times 10^{-8}$ |
| 2,3,4,7,9- | $7.9 \times 10^{-9}$ | $5.8 \times 10^{-9}$ |
| 1,2,3,7,8- | $2.5 \times 10^{-9}$ | $3.1 \times 10^{-9}$ |
| 1,3,4,7,8- | $1.6 \times 10^{-9}$ | $5.8 \times 10^{-9}$ |
| 2,3,4,7,8- | $2.6 \times 10^{-10}$ | $1.3 \times 10^{-10}$ |
| 1,2,4,6,7,8- | $4.2 \times 10^{-8}$ | $2.9 \times 10^{-8}$ |
| 1,2,3,6,7,8- | $1.5 \times 10^{-9}$ | $1.2 \times 10^{-9}$ |
| 2,3,4,6,7,8- | $6.9 \times 10^{-10}$ | $5.8 \times 10^{-10}$ |
| 1,2,3,4,7,8- | $3.6 \times 10^{-10}$ | $3.8 \times 10^{-10}$ |

[a] No data; [b] Position(s) of chlorine atom(s)

affinities ($1.25 \times 10^{-7}$ M and $1.5 \times 10^{-6}$ M, respectively) compared to 2,3,7,8-TCDD ($1.0 \times 10^{-8}$ M), and low agonist activities. Their $EC_{50}$ values for AHH and EROD induction in rat hepatoma cells in culture were greater than $10^{-5}$ M, whereas the $EC_{50}$ values for 2,3,7,8-TCDD were less than $10^{-10}$ M (Keys et al., 1986). These weak Ah receptor agonists have since been investigated as 2,3,7,8-TCDD antagonists (Keys et al., 1986). Hormone antagonists or partial antagonists characteristically block one or more hormone receptor-mediated responses through a variety of mechanisms. Competitive receptor antagonists elicit their effects by interacting with the target receptors and inhibit hormone agonist-mediated responses. This class of receptor antagonists typically exhibit moderate receptor binding affinities and low agonist activities. Several compounds, including 1-amino-3,7,8-trichlorodibenzo-*p*-dioxin, 1,3,6,8-TCDF, 2,4,6,8-TCDF, α-naphthoflavone, 6-methyl-1,3,8-trichlorodibenzofuran (MCDF),

**Table 6b.** Structure activity relationships of polychlorinated dibenzo-*p*-dioxins as inducers of aryl hydrocarbon hydroxylase (AHH) and ethoxyresorufin *O*-deethylase (EROD) in rat hepatoma H-4-II E cells in culture

| Congeners | AHH $EC_{50}$, mol/liter | EROD $EC_{50}$, mol/liter |
|---|---|---|
| 1-[a] | $> 1.0 \times 10^{-4}$ | $> 1.0 \times 10^{-4}$ |
| 2,8- | $> 1.0 \times 10^{-4}$ | $> 1.0 \times 10^{-4}$ |
| 1,2,4- | $4.8 \times 10^{-5}$ | $2.2 \times 10^{-6}$ |
| 2,3,6- | — | — |
| 2,3,7- | $3.6 \times 10^{-7}$ | $1.4 \times 10^{-7}$ |
| 1,2,3,4- | $3.7 \times 10^{-6}$ | $2.4 \times 10^{-6}$ |
| 1,3,7,8- | $5.9 \times 10^{-7}$ | $3.2 \times 10^{-7}$ |
| 2,3,6,7- | $6.1 \times 10^{-8}$ | $1.1 \times 10^{-8}$ |
| 2,3,7,8- | $7.2 \times 10^{-11}$ | $1.9 \times 10^{-10}$ |
| 1,2,3,4,7- | $6.6 \times 10^{-7}$ | $8.2 \times 10^{-7}$ |
| 1,2,4,7,8- | $2.1 \times 10^{-8}$ | $1.1 \times 10^{-8}$ |
| 1,2,3,7,8- | $1.1 \times 10^{-8}$ | $1.7 \times 10^{-8}$ |
| 1,2,3,4,7,8- | $2.1 \times 10^{-9}$ | $4.1 \times 10^{-9}$ |
| 1,2,3,4,5,6,7,8- | $> 1.0 \times 10^{-4}$ | $> 1.0 \times 10^{-4}$ |

[a] Position(s) of chlorine atom(s)

and Aroclor 1254 antagonize the in vitro induction of cytochrome P-450-dependent monooxygenases by 2,3,7,8-TCDD (Keys et al., 1986; Luster et al., 1986; Bannister et al., 1987; Blank et al., 1987).

Table 7 summarizes the comparative agonist activities of 2,3,7,8-TCDD and Aroclor 1254 and illustrates that for several murine Ah receptor-mediated responses including hepatic microsomal AHH/EROD induction, immunotoxicity, and teratogenicity that 2,3,7,8-TCDD is at least $10^5$ times more active than Aroclor 1254 (Bannister et al., 1987; Haake et al., 1987). In contrast, the competitive Ah receptor binding activity of Aroclor 1254 is approximately 500 times lower than that of 2,3,7,8-TCDD (Bandiera et al., 1982). Cotreatment of C57BL/6 mice with an effective dose of 2,3,7,8-TCDD (i.e., 60–90% of a maximum biologic or toxic response) and a range of subeffective doses of Aroclor 1254 resulted in 2,3,7,8-TCDD antagonism which was dependent on the response and the antagonist/agonist ratios (Table 8) (Bannister et al., 1987; Haake et al., 1987; Safe et al., 1988). This PCB mixture represents the first example of a broad spectrum 2,3,7,8-TCDD antagonist for multiple in vivo Ah receptor-mediated responses. Further studies with Aroclor 1254 and individual congeners will be useful as probes for delineating the complex processes which are modulated by 2,3,7,8-TCDD and related compounds.

In the rat, methylchlorodibenzofuran (MCDF) and several related analogues are active as antagonists. At doses up to 200 μmol/kg, MCDF causes no observed induction of rat hepatic microsomal AHH or cytochromes P-4501A1 or P-4501A2 and only a four-fold induction of EROD. Cotreatment of rats with 2,3,7,8-TCDD (16 nmol/kg) plus MCDF (50 μmol/kg) resulted in greater than 50% inhibition of the induction by 2,3,7,8-TCDD of AHH, EROD, and cytochromes P-4501A1 and P-4501A2. Results from double reciprocal plot analyses

**Table 7.** Potencies of 2,3,7,8-tetrachlorodibenzo-*p*-dioxin (TCDD) and Aroclor 1254 in C57BL/6 mice (Bannister et al., 1987; Haake et al., 1987; Safe et al., 1988)

| Effect | 2,3,7,8-TCDD ($ED_{50}$, mol/kg) | Aroclor 1254 ($ED_{50}$, mol/kg) | Relative Potency TCDD/Aroclor 1254 |
|---|---|---|---|
| AHH Induction | $5 \times 10^{-9}$ | $> 5 \times 10^{-4}$ | $> 10^5$ |
| EROD Induction | $7.5 \times 10^{-9}$ | $> 7.5 \times 10^{-4}$ | $> 10^5$ |
| Immunotoxicity (anti-SRBC response) | $2.8 \times 10^{-9}$ | $> 1.5 \times 10^{-4}$ | $> 10^5$ |
| Teratogenicity (cleft palate) | $5 \times 10^{-8}$ | $> 10^{-3}$ | $> 10^5$ |

AHH = aryl hydrocarbon hydroxylase; EROD = ethoxyresorufin *O*-deethylase; SRBC = sheep red blood cells

**Table 8.** Summary: Aroclor 1254 as a 2,3,7,8-tetrachlorodibenzo-*p*-dioxin (TCDD) antagonist in C57BL/6 mice (Bannister et al., 1987; Haake et al., 1987; Safe et al., 1988)

| Response | %Maximum Antagonism | Antagonist/Agonist "Window" |
|---|---|---|
| AHH Induction | 20 | 1667–10,000/1 |
| EROD Induction | 23 | 1667–10,000/1 |
| Immunotoxicity | 100 | 1340–20,160/1 |
| Teratogenicity | > 80 | ± 12,100/1 |
| Thymic Atrophy | 0 | no antagonism |

AHH = aryl hydrocarbon hydroxylase; EROD = ethoxyresorufin *O*-deethylase; ± = approximately

of the receptor binding of [$^3$H]-2,3,7,8-TCDD in the presence of different concentrations of unlabeled antagonists (e.g. Aroclor 1254, MCDF, α-naphthoflavone, and 1-amino-3,7,8-trichlorodibenzo-*p*-dioxin) suggested that these compounds act as competitive antagonists of [$^3$H]-2,3,7,8-TCDD binding. These reports which describe the effects of Ah receptor antagonists further substantiate the proposed receptor-mediated mechanism of action of 2,3,7,8-TCDD and related compounds (Fig. 1).

### *3.2 PCDDs and PCDFs: Structure-Toxicity Relationships*

The in vitro induction assays have been proposed as quantitative short-term test systems to determine the toxic equivalents (e.g. relative to a standard such as 2,3,7,8-TCDD) of PCDDs, PCDFs, and related compounds. Figures 3 and 4 summarize the relative potencies of several PCDD and PCDF congeners to cause either thymic atrophy or body weight loss vs AHH induction in rats (Safe, 1987; Holcomb et al., 1988). There was a positive linear correlation between the $-\log EC_{50}$ values for in vitro AHH induction and the $-\log ED_{50}$ values for

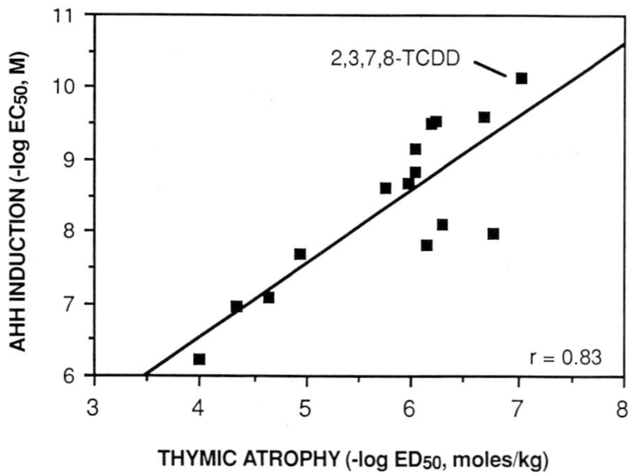


**Fig. 3.** Correlation between the $-\log$ concentration for 50% of maximum response for AHH induction (in vitro) and the $-\log$ dose for 50% of maximum response for thymic atrophy in the rat by several PCDD and PCDF congeners

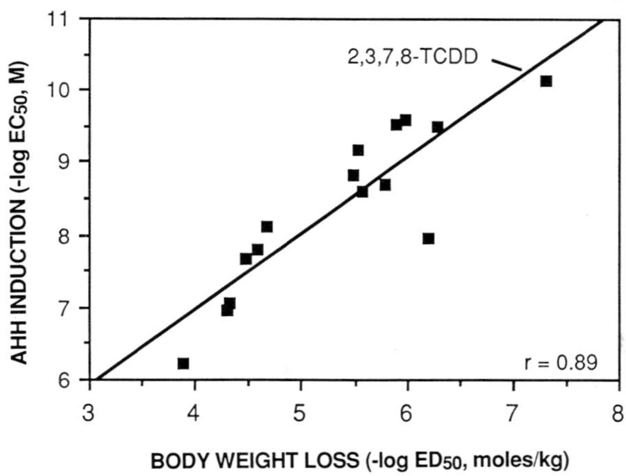


**Fig. 4.** Correlation between the $-\log$ concentration for 50% of maximum response for AHH induction (in vitro) and the $-\log$ dose for 50% of maximum response for body weight loss in the rat by several PCDD and PCDF congeners

thymic atrophy ($r = 0.83$) and body weight loss ($r = 0.89$). The linear correlation between the in vitro and in vivo receptor-mediated responses was also observed when results for several PCDDs and PCBs were included (Safe, 1987). Moreover, for several halogenated aryl hydrocarbons, there was also a linear correlation between their $-\log EC_{50}$ values for AHH induction and their $-\log ED_{50}$ values for immunotoxicity in C57BL/6 mice (Davis and Safe, 1988). Several other studies have reported limited (but comparable) SARs for other proposed Ah receptor-mediated responses including epidermal hyperplasia (Knutson and

Poland, 1982; Poland et al., 1984) and tumor promotion activity (Poland et al., 1982) in HRS/J mice, in vitro keratinization (Knutson and Poland, 1980), and the 2,3,7,8-TCDD-mediated decrease in epidermal growth factor (EGF) receptors (Hudson et al., 1985) and estrogen receptors (Romkes and Safe, 1988).

Limited studies with 2,3,7,8-TCDD antagonists (see Sect. 3.1) have also shown that some of these compounds including MCDF, Aroclor 1254 (see Tables 7 and 8), and α-naphthoflavone antagonize 2,3,7,8-TCDD-mediated toxicity. The structure-toxicity relationships and the identification of antagonists provide additional support for the proposed mechanism of action of PCDDs and PCDFs (Fig. 1). The development and validation of the in vitro induction bioassays will be useful in estimating the potential toxic equivalents of complex mixtures of PCDDs and PCDFs which have been identified as industrial by-products and as environmental contaminants (Safe, 1987).

### *3.3 Genetic Studies*

Pharmacogenetic studies with inbred mice and their backcrosses provided the first evidence which supported the role of the Ah receptor in the induction of AHH (Poland and Glover, 1973, 1974, 1975; Poland et al., 1974). Subsequent research with both responsive and non-responsive strains of mice demonstrate that several toxic effects including hepatotoxicity, immunotoxicity, porphyria, body weight loss, and teratogenicity segregate with the *Ah* locus (Jones and Sweeney, 1980; Poland and Glover, 1980; Vecchi et al., 1980, 1983, Clark et al., 1983; Hassoun et al., 1984; Nagarkatti et al., 1984). It has also been suggested that additional genetic loci may be involved in the hepatotoxicity (Greig et al., 1984) and dermal toxicity of 2,3,7,8-TCDD (Knutson and Poland, 1982; Poland et al., 1982, 1984). Differences in the properties of the Ah receptor from the proposed $Ah^{b-1}$ and $Ah^{b-2}$ responsive strains of mice have been reported (Poland et al., 1986, 1987), and inbreeding experiments indicate that "the expression of these peptides is consistent with codominant inheritance of two alleles at a single locus." Recent gene mapping experiments have indicated that the *Ah* locus, the structural gene which codes for the Ah receptor, is on chromosome 12 (Cobb et al., 1987; Poland et al., 1987). This is in contrast to a previous report which mapped this gene to chromosome 17 (Legraverend et al., 1984).

### *3.4 Models for Studying Ah Receptor-Mediated Responses*

#### Cellular Models

*Dermal Toxicity.* The pleiotropic responses to 2,3,7,8-TCDD and related compounds have been extensively documented, and indirect evidence such as structure-activity relationships and genetic studies support the role of the Ah receptor in many of these processes. However, with few exceptions, the mechanisms associated with these responses are poorly understood. Knutson and Poland (1982) have reported that epidermal hyperplasia, hyperkeratosis, and sebaceous gland metaplasia caused by 2,3,7,8-TCDD in the skin of HRS/J mice bearing the recessive mutation, hairless (*hr*/*hr*), are Ah receptor-mediated responses. The Ah receptor has been identified in the epidermis of both HRS/J hairless (*hr*/*hr*) and haired (*hr*/ ± ) mice, and 2,3,7,8-TCDD induces epidermal AHH in these strains of mice. In contrast, the other epidermal responses were observed only in the hairless mice, and it was suggested that an interaction of both the *Ah* and *hr* genetic loci are required for development of the proliferative/metaplastic skin response. Molloy and coworkers (1987) also reported that 2,3,7,8-TCDD elicited in haired and hairless HRS/J mice a dose-dependent epidermal hyperplasia which was associated with altered keratin expression. In both strains of mice, 2,3,7,8-TCDD caused an increase in keratins normally observed in proliferating basal cells and primary epidermal cell cultures, whereas only in the hairless mice was a decrease in the synthesis of specific keratins associated with suprabasal differentiation observed.

Treatment of human epidermal keratinocytes in culture with 2,3,7,8-TCDD increased the number of envelope competent and terminally differentiated cells, decreased the number of proliferating basal cells, and decreased EGF receptor binding (Greenlee et al., 1985a; Osborne and Greenlee, 1985). This group has suggested that 2,3,7,8-TCDD acts on proliferating basal cells by modulating their responsiveness to growth factors such as EGF and that this model should provide additional insights on the molecular mechanism involved in 2,3,7,8-TCDD-mediated changes in growth factors and their receptors and the cellular changes which occur in response to these effects.

*Thymic Toxicity*. Mechanistic studies of the action of 2,3,7,8-TCDD on the thymus have established that thymic epithelial cells in culture are a target for toxic halogenated aromatics (Greenlee et al., 1985b; Cook et al., 1987). Thymic epithelial (TE) cells isolated from C57BL/6 mice or human thymuses contain the Ah receptor, and 2,3,7,8-TCDD characteristically induces cytochrome P-450-dependent monooxygenases in these cultured cells. Moreover, 2,3,7,8-TCDD and related compounds stereoselectively suppress the thymic epithelial-dependent maturation of thymocytes in the in vitro model system. The development of this type of in vitro model system coupled with molecular biology approaches will also be useful tools for further delineating the mechanisms of action of PCDDs, PCDFs, and related compounds as modulators of immune responses.

## Molecular Models

Models which seek to explain the mechanism of action 2,3,7,8-TCDD must account for the diversity of responses elicited by this compound. One idea is that

2,3,7,8-TCDD alters the expression of a set of genes in each 2,3,7,8-TCDD-responsive cell type by a mechanism that involves the Ah receptor. One convenient model reponse for studying 2,3,7,8-TCDD action at the molecular level is the induction of AHH activity. This enzyme activity is catalyzed by the cytochrome P-4501A1 isozyme; most detailed knowledge of 2,3,7,8-TCDD action has come from the study of 2,3,7,8-TCDD-responsive cytochrome P-450 genes (Nebert and Gonzalez, 1987; Whitlock, 1987). It is assumed that 2,3,7,8-TCDD acts at other genes by a similar mechanism. If so, then the study of AHH induction by 2,3,7,8-TCDD may provide insight into the overall mechanism of action of PCDDs and PCDFs.

Nuclear transcription studies indicate that 2,3,7,8-TCDD activates the rate of cytochrome P-450 gene transcription (Gonzalez et al., 1984; Israel and Whitlock, 1984). The increased transcription rate accounts for the subsequent rise in cytochrome P-450 mRNA accumulation and AHH activity. In cells that contain defective Ah receptors, 2,3,7,8-TCDD fails to activate gene transcription or does so poorly (Israel and Whitlock, 1983, 1984). The findings to date imply that the activation of gene transcription by 2,3,7,8-TCDD requires both the formation of 2,3,7,8-TCDD-receptor complex and an interaction between the complex and a component of the cell nucleus. The generation of a functional 2,3,7,8-TCDD-receptor complex requires a temperature-dependent "activation" step(s) (Okey et al., 1980). Activation has the effect of increasing the affinity of the 2,3,7,8-TCDD-receptor complex for nuclear binding sites, presumably on chromatin (Okey et al., 1979; Whitlock and Galeazzi, 1984). The biochemical event(s) that results in receptor activation is not known.

The interaction of the activated 2,3,7,8-TCDD-receptor complex with the nucleus leads rapidly to an increase in gene transcription (Israel and Whitlock, 1984); furthermore, the response does not require ongoing protein synthesis (Israel et al., 1985). These findings imply that the 2,3,7,8-TCDD-receptor complex can activate gene transcription directly, without a requirement for intervening biochemical events, such as the generation of "second messengers" or the induction of other proteins (Knutson and Poland, 1984a,b).

The 2,3,7,8-TCDD-receptor complex acts at a genomic "switch" to activate cytochrome P-450 gene transcription (Gonzalez and Nebert, 1985; Jones et al., 1985; Sogawa et al., 1986). Recombinant DNA and gene transfer techniques have been used to isolate from the 5′ upstream region of the cytochrome P-450 gene, a segment of DNA that can confer responsiveness to 2,3,7,8-TCDD upon a different gene. This genomic regulatory region, which has been termed a "dioxin-responsive element" (DRE), functions as a transcriptional enhancer and requires the 2,3,7,8-TCDD-receptor complex for its action, because it fails to function in receptor-defective cells (Jones et al., 1986). In addition, the DRE of the mouse can function in human cells, and the DREs of the rat and human can function in mouse cells (Jones et al., 1986; Sogawa et al., 1986; Fujisawa-Sehara et al., 1987). These findings imply that the mechanism that confers 2,3,7,8-TCDD-responsiveness upon a gene is functionally similar in rodent and human cells. Molecular biological techniques have been used to show that the 2,3,7,8-TCDD-receptor complex specifically interacts with the DRE, both in the intact nucleus of the cell (Durrin and Whitlock, 1987) and when purified DNA and nuclear proteins are studied (Denison et al., 1988). The 2,3,7,8-TCDD-receptor complex recognizes a specific DNA sequence within the DRE, and the binding of the complex to this "receptor recognition sequence" appears necessary for the subsequent activation of gene transcription (personal communication from J.P. Whitlock, Jr., Stanford University, Stanford, CA). However, it is not yet clear whether this protein-DNA interaction is sufficint to elicit a transcriptional response.

The proposed mechanism of 2,3,7,8-TCDD action (and, by inference, that of other PCDDs and PCDFs) involves the specific binding of the ligand to an intracellular protein receptor, followed by the specific binding of the 2,3,7,8-TCDD-receptor complex to a particular DNA recognition sequence. In principle, both of these events are reversible, because they involve non-covalent biochemical interactions. The mechanism by which the interaction between the 2,3,7,8-TCDD-receptor complex and its recognition sequence in DNA leads to a change in gene expression is not known. Studies in other systems indicate that multiple components are involved in the regulation of gene transcription, so it seems likely that this situation will also be true for genes that respond to PCDDs and PCDFs.

## 4 Conclusions

The available evidence indicates that 2,3,7,8-TCDD does not act via direct damage to DNA. Biochemical and genetic evidence from experimental studies suggests that 2,3,7,8-TCDD acts by binding reversibly, saturably, and with high affinity to an intracellular protein, designated the Ah receptor. The binding of the ligand (PCDD-PCDF)-receptor complex to nuclear material modulates gene expression. This may occur in a tissue- and species-specific fashion. The subsequent events leading to the observed biological responses, such as chloracne and tumor promotion, are not clear. The evidence suggests that the structure of the Ah receptor is heterogeneous among species. However, it is not known if these structural differences influence species susceptibility to PCDDs and PCDFs. The Ah receptor is necessary but not sufficient for eliciting some of the biologic and toxic responses caused by PCDDs and PCDFs. Therefore, other factors are involved in these processes. The receptor-mediated model implies that early events in 2,3,7,8-TCDD action (i.e., occupied receptor-DNA interactions) are reversible and it is clear that some of the biologic and toxic effects are also reversible. However, not all of the effects of 2,3,7,8-TCDD and related compounds are reversible.

# 5 References

Bandiera S, Safe S, Okey AB (1982) Binding of polychlorinated biphenyls classified as either phenobarbitone-, 3-methylcholanthrene- or mixed-type inducers to cytosolic Ah receptor. Chem. Biol. Interact. 39: 259–277

Bandiera S, Sawyer TW, Campbell MA, Fujita T, Safe S (1983) Competitive binding to the cytosolic 2,3,7,8-TCDD receptor. Effects of structure on the affinities of substituted halogenated biphenyls: a QSAR analysis. Biochem. Pharmacol. 32: 3803–3813

Bandiera S, Sawyer T, Romkes M, Zmudzka B, Safe L, Mason G, Keys B, Safe S (1984) Polychlorinated dibenzofurans (PCDFs): effects of structure on binding to the 2,3,7,8-TCDD cytosolic receptor protein, AHH induction and toxicity. Toxicology 32: 131–144

Bannister R, Safe S (1987) Synergistic interactions of 2,3,7,8-TCDD and 2,2′,4,4′,5,5′-hexachlorobiphenyl in C57BL/6J and DBA/2J mice: role of the Ah receptor. Toxicology 44: 159–169

Bannister R, Davis D, Zacharewski T, Tizard I, Safe S (1987) Aroclor 1254 as a 2,3,7,8-tetrachlorodibenzo-*p*-dioxin antagonist: effects on enzyme induction and immunotoxicity. Toxicology 46: 29–42

Bastomasky CH (1977) Enhanced thyroxine metabolism and high uptake goiters in rats after a single dose of 2,3,7,8-tetrachlorodibenzo-*p*-dioxin. Endocrinology 101: 292–296

Beatty PW, Neal RA (1976) Induction of DT-diaphorase by 2,3,7,8-tetrachlorodibenzo-*p*-dioxin (TCDD). Biochem. Biophys. Res. Commun. 68: 197–204

Birnbaum LS, Weber H, Harris MW, Lamb JC, McKinney JD (1985) Toxic interaction of specific polychlorinated biphenyls and 2,3,7,8-tetrachlorodibenzo-*p*-dioxin: increased incidence of cleft palate in mice. Toxicol. Appl. Pharmacol. 77: 292–302

Birnbaum LS, Harris MW, Barnhart ER, Morrissey RE (1987) Teratogenicity of three polychlorinated dibenzofurans in C57BL/6N mice. Toxicol. Appl. Pharmacol. 90: 206–216

Blank JA, Tucker AN, Sweatlock J, Gasiewicz TA, Luster MI (1987) alpha-Naphthoflavone antagonism of 2,3,7,8-tetrachlorodibenzo-*p*-dioxin induced murine lymphocyte ethoxyresorufin *O*-deethylase activity and immunosuppression. Mol. Pharmacol. 32: 169–172

Carlstedt-Duke JM (1979) Tissue distribution of the receptor for 2,3,7,8-tetrachlorodibenzo-*p*-dioxin in the rat. Cancer Res. 39: 3172–3176

Carlstedt-Duke JM, Elfstrom G, Hogberg B, Gustafsson JA (1979) Ontogeny of the rat hepatic receptor for 2,3,7,8-tetrachlorodibenzo-*p*-dioxin and its endocrine independence. Cancer Res. 39: 4653–4656

Carlstedt-Duke JM, Harnemo UB, Hogberg B, Gustafsson JA (1981) Interaction of the hepatic receptor protein for 2,3,7,8-tetrachlorodibenzo-*p*-dioxin with DNA. Biochim. Biophys. Acta 672: 131–141

Chapman DE, Schiller CM (1985) Dose-related effects of 2,3,7,8-tetrachlorodibenzo-*p*-dioxin (TCDD) in C57BL/6J and DBA/2J mice. Toxicol. Appl. Pharmacol. 78: 147–157

Clark DA, Sweeney G, Safe S, Hancock E, Kilburn DG, Gauldie J (1983) Cellular and genetic basis for suppression of cytotoxic T cell generation by haloaromatic hydrocarbons. Immunopharmacology 6: 143–153

Cobb RR, Stoming TA, Whitney JB III (1987) The aryl hydrocarbon hydroxylase (*Ah*) locus and a novel restriction-fragment length polymorphism (RFLP) are located on mouse chromosome 12. Biochem. Genet. 25: 401–413

Cook JC, Dold KM, Greenlee WF (1987) An in vitro model for studying the toxicity of 2,3,7,8-tetrachlorodibenzo-*p*-dioxin to human thymus. Toxicol. Appl. Pharmacol. 89: 256–268

Cuthill S, Poellinger L, Gustafsson JA (1987) The receptor for 2,3,7,8-tetrachlorodibenzo-*p*-dioxin in the mouse hepatoma cell line Hepa 1c1c7. J. Biol. Chem. 262: 3477–3481

Davis D, Safe S (1988) Immunosuppressive activities of polychlorinated dibenzofuran congeners: quantitative structure activity relationships and interactive effects. Toxicol. Appl. Pharmacol. 94: 141–149

Deitrich RA, Bludeau P, Stock T, Roper M (1977) Induction of different rat liver supernatant aldehyde dehydrogenases by phenobarbital and tetrachlorodibenzo-*p*-dioxin. J. Biol. Chem. 252: 6169–6176

Denison MS, Vella LM, Okey AB (1986a) Structure and function of the Ah receptor for 2,3,7,8-tetrachlorodibenzo-*p*-dioxin: species differences in molecular properties of the receptors from mouse and rat hepatic cytosols. J. Biol. Chem. 261: 3987–3995

Dension MS, Vella LM, Okey AB (1986b) Hepatic Ah receptor for 2,3,7,8-tetrachlorodibenzo-*p*-dioxin: partial stabilization by molybdate. J. Biol. Chem. 261: 10189–10195

Denison MS, Wilkinson CF, Okey AB (1986c) Ah receptor for 2,3,7,8-tetrachlorodibenzo-*p*-dioxin: comparative studies in mammalian and non-mammalian species. Chemosphere 15: 1672–1675

Denison MS, Okey AB, Hamilton JW, Bloom SE, Wilkinson CF (1986d) Ah receptor for 2,3,7,8-tetrachlorodibenzo-*p*-dioxin: ontogeny in chick embryo liver. J. Biochem. Toxicol. 1: 39–46

Denison MS, Harper PA, Okey AB (1986e) Ah receptor for 2,3,7,8-tetrachlorodibenzo-*p*-dioxin: codistribution of unoccupied receptor with cytosolic marker enzymes during fractionation of mouse liver, rat liver, and cultured Hepa-1c1 cells. Eur. J. Biochem. 155: 223–229

Denison MS, Vella LM and Okey AB (1987) Structure and function of the Ah receptor: sulfhydryl groups required for binding of 2,3,7,8-tetrachlorodibenzo-*p*-dioxin to cytosolic receptor from rodent livers. Arch. Biochem. Biophys. 252: 388–395

Denison MS, Fisher JM, Whitlock JP Jr (1988) Inducible receptor-dependent protein-DNA interactions at a dioxin-responsive transcriptional enhancer. Proc. Natl. Acad. Sci. USA 85: 2528–2532

Denomme MA, Homonko K, Fujita T, Sawyer T, Safe S (1985) The effects of substituents on the cytosolic receptor binding avidities and aryl hydrocarbon hydroxylase induction potencies of 7-substituted-2,3-dichlorodibenzo-*p*-dioxins: a quantitative structure-activity relationship analysis. Mol. Pharmacol. 27: 656–661

Denomme MA, Homonko K, Fujita T, Sawyer T, Safe S (1986a) Substituted polychlorinated dibenzofuran receptor binding affinities and aryl hydrocarbon hydroxylase induction potencies: a QSAR analysis. Chem. Biol. Interact. 57: 175–187

Denomme MA, Leece B, Li A, Towner R, Safe S (1986b) Elevation of 2,3,7,8-TCDD polychlorinated biphenyls: structure-activity relationships. Biochem. Pharmacol. 35: 277–282

Durrin LK, Whitlock JP Jr (1987) In situ protein-DNA interactions at a dioxin-responsive enhancer associated with the cytochrome P1-450 gene. Molec. Cell Biol. 7: 3008–3011

Farrell K, Safe S (1987) Absence of positive cooperativity in the binding of 2,3,7,8-tetrachlorodibenzo-*p*-dioxin to its cytosolic receptor protein. Biochem. J. 244: 539–546

Farrell K, Safe L, Safe S (1987) Synthesis and aryl hydrocarbon receptor binding properties of radiolabeled polychlorinated dibenzofuran congeners. Arch. Biochem. Biophys. 259: 185–195

Fick DF, Firestone D, Ress J, Allen JR (1973) Studies of the chick edema disease. 10. Toxicity of chick edema factors in the chick, chick embryo and monkey. Poult. Sci. 52: 1637–1641

Fujisawa-Sehara A, Sogawa K, Yamane M, Fujii-Kuriyama Y (1987) Characterization of xenobiotic responsive elements upstream from the drug-metabolizing cytochrome P-450c gene: a similarity to glucocorticoid regulatory elements. Nucleic Acids Res. 15: 4179–4191

Gasiewicz TA, Bauman PA (1987) Heterogeneity of the rat hepatic Ah receptor and evidence for transformation in vitro and in vivo. J. Biol. Chem. 262: 2116–2120

Gasiewicz TA, Rucci G (1984a) Cytosolic receptor for 2,3,7,8-tetrachlorodibenzo-*p*-dioxin: evidence for a homologous nature among various mammalian species. Mol. Pharmacol. 26: 90–98

Gasiewicz TA, Rucci G (1984b) Examination and rapid analysis of hepatic cytosolic receptors for 2,3,7,8-tetrachlorodibenzo-*p*-dioxin using gel-permeation high performance liquid chromatography. Biochim. Biophys. Acta 798: 37–45

Gasiewicz TA, Rucci G (1987) Long-term kinetics of the Ah receptor in hamsters administered [3*H*]-2,3,7,8-tetrachlorodibenzo-*p*-dioxin (TCDD). Toxicologist 7: A647 [Abstract]

Gasiewicz TA, Olson JR, Geiger LH, Neal RA (1983) Absorption, distribution, and metabolism of 2,3,7,8-tetrachlorodibenzo-*p*-dioxin (TCDD) in experimental animals. In: Tucker RE, Young AL and Gray AP (eds) Human and environmental risks of chlorinated dioxins and related compounds. Plenum, New York, pp. 495–525

Gasiewicz TA, Ness WC, Rucci G (1984) Ontogeny of the cytosolic receptor for 2,3,7,8-tetrachlorodibenzo-*p*-dioxin in rat liver, lung and thymus. Biochem. Biophys. Res. Commun. 118: 183–190

Gasiewicz TA, Rucci G, Henry EC, Baggs RB (1986) Changes in hamster hepatic cytochrome P-450, ethoxycoumarin *O*-deethylase, and reduced NAD(P): menadione oxidoreductase following treatment with 2,3,7,8-tetrachlorodibenzo-*p*-dioxin. Biochem. Pharmacol. 35: 2737–2742

Geiger LE, Neal RA (1981) Mutagenicity testing of 2,3,7,8-tetrachlorodibenzo-*p*-dioxin in histidine auxotrophs of *Salmonella typhimurium*. Toxicol. Appl. Pharmacol. 59: 125–129

Gillner M, Bergman J, Cambillau C, Fernstrom B, Gustafsson JA (1985) Interactions of indoles with specific binding sites for 2,3,7,8-tetrachlorodibenzo-*p*-dioxin in rat liver. Mol. Pharmacol. 28: 357–363

Giri AK (1986) Mutagenic and genotoxic effects of 2,3,7,8-tetrachlorodibenzo-*p*-dioxin: a review. Mutat. Res. 168: 241–248

Gonzalez FJ, Nebert DW (1985) Autoregulation plus upstream positive and negative control regions associated with transcriptional activation of the mouse cytochrome P1-450 gene. Nucleic Acids Res. 13: 7269–7288

Gonzalez FJ, Tukey RH, Nebert DW (1984) Structural gene products of the *Ah* locus: transcriptional regulation of cytochrome P1-450 and P3-450 mRNA levels by 3-methylcholanthrene. Mol. Pharmacol. 26: 117–121

Gottlicher M, Cikryt P (1987) Induction of the aromatic hydrocarbon receptor by *trans*-4-acetylaminostilbene in rat liver: comparison with other aromatic amines. Carcinogenesis 8: 1021–1023

Green S, Moreland F, Sheu C (1977) Cytogenetic effects of rat bone marrow cells. FDA By-Lines 6: 292–294

Greenlee WF, Poland A (1978) An improved assay of 7-ethoxycoumarin *O*-deethylase activity: induction of hepatic enzyme activity in C57BL/6J and DBA/2J mice by phenobarbital, 3-methylcholanthrene and 2,3,7,8-tetrachlorodibenzo-*p*-dioxin. J. Pharmacol. Exp. Ther. 205: 596–605

Greenlee WF, Dold KM, Osborne R (1985a) Actions of 2,3,7,8-tetrachlorodibenzo-*p*-dioxin (TCDD) on human epidermal keratinocytes in culture. In Vitro Cell Dev. Biol. 21: 509–512

Greenlee WF, Dold KM, Irons RD and Osborne R (1985b) Evidence for direct action of 2,3,7,8-tetrachlorodibenzo-*p*-dioxin (TCDD) on thymic epithelium. Toxicol. Appl. Pharmacol. 79: 112–120

Greig JB, Francis JE, Kay SJ, Lovell DP, Smith AG (1984) Incomplete correlation of 2,3,7,8-tetrachlorodibenzo-*p*-dioxin hepatotoxicity with Ah phenotype in mice. Toxicol. Appl. Pharmacol. 74: 17–25

Gudas JM, Karenlampi SO, Hankinson O (1986) Intracellular location of the Ah receptor. J. Cell. Physiol. 128: 441–448

Gupta BW, McConnell EE, Goldstein JA, Harris MW, Moore JA (1983) Effects of a polybrominated biphenyl in the rat and mouse. I. Six-month exposure. Toxicol. Appl. Pharmacol. 68: 1–18

Haake JM, Safe S, Mayura K, Phillips TD (1987) Aroclor 1254 as an antagonist of the teratogenicity of 2,3,7,8-tetrachlorodibenzo-*p*-dioxin. Toxicol. Lett. 38: 299–306

Hankinson O (1979) Single-step selection of clones of a mouse hepatoma line deficient in aryl hydrocarbon hydroxylase. Proc. Natl. Acad. Sci. [USA] 76: 373–376

Hannah RR (1988) DNA binding forms of rat Ah receptor complex. Toxicologist 8: A139 [Abstract]

Hannah RR, Nebert DW, Eisen HJ (1981) Regulatory gene product of the Ah complex: comparison of 2,3,7,8-tetrachlorodibenzo-*p*-dioxin and 3-methyl-cholanthrene binding to several moieties in mouse liver cytosol. J. Biol. Chem. 256: 4584–4590

Hannah RR, Lund J, Poellinger L, Gillner M, Gustafsson JA (1986) Characterization of the DNA-binding properties of the receptor for 2,3,7,8-tetrachlorodibenzo-*p*-dioxin. Eur. J. Biochem. 156: 237–242

Hassoun E, d'Argy R, Dencker L, Lundin LG, Borwell P (1984) Teratogenicity of 2,3,7,8-tetrachlorodibenzofuran in BXD recombinant inbred strains. Toxicol. Lett. 23: 37–42

Henry EC, Gasiewicz TA (1987) Changes in thyroid hormones and thyroxine glucuronidation in hamsters compared with rats following treatment with 2,3,7,8-tetrachlorodibenzo-*p*-dioxin. Toxicol. Appl. Pharmacol. 89: 165–174

Henry EC, Kester JE, Gasiewicz TA (1988) Effects of SH-modifying reagents on the rat hepatic Ah receptor: inhibition of ligand binding and transformation, disruption of the ligand-receptor complex. Biochem. Biophys. Acta 964: 361–376

Holcomb M, Yao C, Safe S (1988) The biologic and toxic effects of polychlorinated dibenzo-*p*-dioxins and dibenzofurans in the guinea pig: quantitative structure-activity relationships. Biochem. Pharmacol. 37: 1535–1539

Hong R, Taylor K, Abonour R (1987) Immune abnormalities associated with chronic TCDD exposure in rhesus. Presented at Dioxin '87: Seventh International Symposium on Chlorinated Dioxins and Related Compounds, October 4–9, Las Vegas. [Abstract]. Available from: [University of Nevada, Las Vegas]

Hudson LG, Toscano WA Jr, Greenlee WF (1985) Regulation of epidermal growth factor binding in a human keratinocyte cell line by 2,3,7,8-tetrachlorodibenzo-*p*-dioxin. Toxicol. Appl. Pharmacol. 77: 251–259

Israel DI, Whitlock JP Jr (1983) Induction of mRNA specific for cytochrome P1-450 in wild type and variant mouse hepatoma cells. J. Biol. Chem. 258: 10390–10394

Israel DI, Whitlock JP Jr (1984) Regulation of cytochrome P1-450 gene transcription by 2,3,7,8-tetrachlorodibenzo-*p*-dioxin in wild type and variant mouse hepatoma cells. J. Biol. Chem. 259: 5400–5402

Israel DI, Estolano MG, Galeazzi DR, Whitlock JP Jr (1985) Superinduction of cytochrome P1-450 gene transcription by inhibition of protein synthesis in wild type and variant mouse hepatoma cells. J. Biol. Chem. 260: 5648–5653

Jackson WT (1972) Regulation of mitosis. II. Cytological effects of 2,4,5-trichlorophenoxy acetic acid and of dioxin contaminants in 2,4,5-T formulations. J. Cell Sci. 10: 15–25

Jones KG, Sweeney GD (1980) Dependence of the porphyrogenic effect of 2,3,7,8-tetrachlorodibenzo-*p*-dioxin upon inheritance of aryl hydrocarbon hydroxylase responsiveness. Toxicol. Appl. Pharmacol. 53: 42–49

Jones PB, Galeazzi DR, Fisher JM, Whitlock JP Jr (1985) Control of cytochrome P1-450 gene expression by dioxin. Science 227: 1499–1502

Jones PB, Durrin LK, Galeazzi DR, Whitlock JP (1986) Control of cytochrome P1-450 gene expression: analysis of a dioxin-responsive enhancer system. Proc. Natl. Acad. Sci. [USA] 83: 2802–2806

Kahl GF, Friederici DE, Bigelow SW, Okey AB, Nebert DW (1980) Ontogenic expression of regulatory and structural gene products associated with the *Ah* locus: comparison of rat, mouse, rabbit, and Sigmoden hispedis. Dev. Pharmacol. Ther. 1: 137–162

Karenlampi SO, Eisen HJ, Hankinson O, Nebert DW (1983) Effects of cytochrome P1-450 inducers on the cell-surface receptors for epidermal growth factor, phorbol 12,13-dibutyrate or insulin of cultured mouse hepatoma cells. J. Biol. Chem. 258: 10378–10383

Kelling CK, Menahan LA, Peterson RE (1987) Hepatic indices of thyroid status in rats treated with 2,3,7,8-tetrachlorodibenzo-*p*-dioxin. Biochem. Pharmacol. 36: 283–291

Kester JE, Gasiewicz TA (1987) Characterization of the in vitro stability of the rat hepatic receptor for 2,3,7,8-tetrachlorodibenzo-*p*-dioxin (TCDD). Arch. Biochem. Biophys. 252: 606–625

Keys B, Piskorska-Pliszczynska J, Safe S (1986) Polychlorinated dibenzofurans as 2,3,7,8-TCDD antagonists: in vitro inhibition of monooxygenase enzyme induction. Toxicol. Lett. 31: 151–158

Kirsch R, Fleischner G, Kamisaka K, Arias IM (1975) Structural and functional studies of ligandin, a major renal organic anion-binding protein. J. Clin. Invest. 55: 1009–1019

Knutson JC, Poland A (1980) Keratinization of mouse teratoma cell line XB produced by 2,3,7,8-tetrachlorodibenzo-*p*-dioxin: an in vitro model of toxicity. Cell 22: 27–36

Knutson JC, Poland A (1982) Response of murine epidermis to 2,3,7,8-tetrachlorodibenzo-*p*-dioxin: interaction of *Ah* and *hr* loci. Cell 30: 225–234.

Knutson JC, Poland A (1984a) 2,3,7,8-Tetrachlorodibenzo-*p*-dioxin: examination of the biochemical effects involved in the proliferation and differentiation of XB cells. J. Cell. Physiol. 121: 143–151

Knutson JC, Poland A (1984b) XB cells: an in vitro model for the differentiation and proliferation response to 2,3,7,8-TCDD. In: Poland A and Kimbrough RD (eds) Banbury Report 18. Biological mechanisms of dioxin action. Cold Spring Harbor, NY: Cold Spring Harbor Laboratory. pp 381–389

Kociba R (1984) Evaluation of the carcinogenic and mutagenic potential of 2,3,7,8-TCDD and other chlorinated dioxins. In: Poland A and Kimbrough RD (eds) Banbury Report 18. Biological mechanisms of dioxin action. Cold Spring Harbor, NY: Cold Spring Harbor Laboratory. pp 73–84

Kociba RJ, Schwetz BA (1982a) Toxicity of 2,3,7,8-tetrachlorodibenzo-*p*-dioxin (TCDD). Drug Metab. Rev. 13: 387–406

Kociba RJ, Schwetz BA (1982b) A review of the toxicity of 2,3,7,8-tetrachlorodibenzo-*p*-dioxin (TCDD) with a comparison of the toxicity of the other chlorinated dioxin isomers. Assoc. Food Drug Officials Q. Bull. 46: 168–198

Kumaki K, Jensen NM, Shire JG, Nebert DW (1977) Genetic differences in induction of cytosol reduced NAD(P): menadione oxidoreductase and microsomal aryl hydrocarbon hydroxylase in the mouse. J. Biol. Chem. 252: 157–165

Lamb JC IV, Harris MW, McKinney JD, Birnbaum LS (1986) Effects of thyroid hormones on the induction of cleft palate by 2,3,7,8-tetrachlorodibenzo-*p*-dioxin (TCDD) in C57BL/6N mice. Toxicol. Appl. Pharmacol. 84: 115–124

Leece B, Denomme MA, Towner R, Li AMA, Landers J, Safe S (1987) Nonadditive interactive effects of polychlorinated biphenyl congeners in rats: role of the 2,3,7,8-tetrachlorodibenzo-*p*-dioxin receptor. Can. J. Physiol. Pharmacol. 65: 1908–1912

Legraverend C, Hannah RR, Eisen HJ, Owens IS, Nebert DW, Hankinson O (1982) Regulatory gene product of the *Ah* locus: characterization of receptor mutants among mouse hepatoma clones. J. Biol. Chem. 257: 6402–6407

Legraverend C, Karenlampi SO, Bigelow SW, Lalley PA, Kozak CA, Womack JE, Nebert DW (1984) Aryl hydrocarbon hydroxylase induction by benzo(a)anthracene: regulatory gene localized to the distal portion of mouse chromosome 17. Genetics 107: 447–461

Lesca P, Fernandez N, Roy M (1987) The binding components for 2,3,7,8-tetrachlorodibenzo-*p*-dioxin and polycyclic aromatic hydrocarbons: separation from the rat and mouse hepatic cytosol and characterization of a light density component. J. Biol. Chem. 262: 4827–4835

Long G, McKinney J, Pedersen L (1987) Polychlorinated dibenzofuran (PCDF) binding to the Ah receptor(s) and associated enzyme induction: theoretical model based on molecular parameters. Quant. Struct. Act. Relat. 6: 1–7

Loprieno N, Sbrana I, Rusciano O, Lascialfari D, Lari T (1982) In vivo cytogenetic studies on mice and rats exposed to 2,3,7,8-tetrachlorodibenzo-*p*-dioxin. In: Hutzinger O, Frei RW, Merian E and Pocchiari P (eds) Chlorinated dioxins and related compounds: impact on the environment. Pergamon, Elmsford, NY, pp 419–428

Lubet RA, Connolly G, Kouri RE, Nebert DW, Bigelow SW (1983) Biological effects of the Sudan dyes: role of the Ah cytosolic receptor. Biochem. Pharmacol. 32: 3053–3058

Luster MI, Hong LH, Boorman GA, Clark G, Hayes HT, Greenlee WF, Dold K, Tucker AN (1985) Acute myelotoxic responses in mice exposed to 2,3,7,8-tetrachlorodibenzo-*p*-dioxin (TCDD). Toxicol. Appl. Pharmacol. 81: 156–165

Luster MI, Hong LH, Osborne R, Blank JA, Clark G, Silver MT, Boorman GA, Greenlee WF (1986) 1-Amino-3,7,8-trichlorodibenzo-*p*-dioxin: a specific antagonist for TCDD-induced myelotoxicity. Biochem. Biophys. Res. Commun. 139: 747–756

Madhukar BV, Brewster DW, Matsumura F (1984) Effects of in vivo administered 2,3,7,8-tetrachlorodibenzo-*p*-dioxin on receptor binding of epidermal growth factor in the hepatic plasma membrane of rat, guinea pig, mouse and hamster. Proc. Natl. Acad. Sci. [USA] 81: 7407–7411

Manchester DK, Gordon SK, Golas CL, Roberts EA, Okey AB (1987) Ah receptor in human placenta: stabilization by molybdate and characterization of binding of 2,3,7,8-tetrachlorodibenzo-*p*-dioxin, 3-methylcholanthrene, and benzo[a]pyrene. Cancer Res. 47: 4861–4868

Mason ME, Okey AB (1982) Cytosolic and nuclear binding of 2,3,7,8-tetrachlorodibenzo-*p*-dioxin to the Ah receptor in extra-hepatic tissues of rats and mice. Eur. J. Biochem. 123: 209–215

Mason G, Safe S (1986) Synthesis, biologic and toxic effects of the major 2,3,7,8-tetrachlorodibenzo-*p*-dioxin metabolites in the rat. Toxicology 41: 153–159

Mason G, Sawyer T, Keys B, Bandiera S, Romkes M, Piskorska-Pliszczynska J, Zmudzka B, Safe S (1985) Polychlorinated dibenzofurans (PCDFs): correlation between in vivo and in vitro structure-activity relationships. Toxicology 37: 1–12

Mason G, Farrell K, Keys B, Piskorska-Pliszczynska J, Safe L, Safe S (1986) Polychlorinated dibenzo-*p*-dioxins: quantitative in vitro and in vivo structure-activity relationships. Toxicology 41: 21–31

Mason G, Zacharewski T, Denomme MA, Safe L, Safe S (1987) Polybrominated dibenzo-*p*-dioxins and related compounds: quantitative in vivo and in vitro structure-activity relationships. Toxicology 44: 245–255

Mebus CA, Reddy VR, Piper WN (1987) Depression of rat testicular 17-hydroxylase and 17,20-lyase after administration of 2,3,7,8-tetrachlorodibenzo-*p*-dioxin (TCDD). Biochem. Pharmacol. 36: 727–731.

Meyne J, Allison DC, Bose K, Jordan SW, Ridolpho PF, Smith J (1985) Hepatotoxic doses of dioxin do not damage mouse bone marrow chromosomes. Mutat. Res. 157: 63–69

Miller AG, Whitlock JP Jr (1981) Novel variants in benzo(a)pyrene metabolism: isolation by fluorescence-activated cell sorting. J. Biol. Chem. 256: 2433–2437.

Miller AG, Israel D, Whitlock JP Jr (1983) Biochemical and genetic analysis of variant mouse hepatoma cells defective in the induction of benzo(a)pyrene-metabolizing enzyme activity. J. Biol. Chem. 258: 3523–3527

Molloy CJ, Gallo MA and Laskin JD (1987) Alterations in the expression of specific epidermal keratin markers in the hairless mouse by the topical application of the tumor promoters 2,3,7,8-tetrachlorodibenzo-*p*-dioxin and the phorbol ester 12-*O*-tetradecanoylphorbol-13-acetate. Carcinogenesis 8: 1193–1199

Moore RW, Potter CL, Theobald HM, Robinson JA, Peterson RE (1985) Androgenic deficiency in male rats treated with 2,3,7,8-tetrachlorodibenzo-*p*-dioxin. Toxicol. Appl. Pharmacol. 79: 99–111

Moore RW, Kleeman JM, Peterson RE (1987) Inhibition of testosterone production in tests from 2,3,7,8-tetrachlorodibenzo-*p*-dioxin (TCDD)-treated rats. Toxicologist 7: 504 [Abstract]

Nagarkatti PS, Sweeney GD, Gauldie J, Clark DA (1984) Sensitivity to suppression of cytotoxic T cell generation by 2,3,7,8-tetrachlorodibenzo-*p*-dioxin (TCDD) is dependent on the *Ah* genotype of the murine host. Toxicol. Appl. Pharmacol. 72: 169–176

Neal RA, Beatty PW, Gasiewicz TA (1979) Studies on the mechanisms of toxicity of 2,3,7,8-tetrachlorodibenzo-*p*-dioxin (TCDD). Ann. N.Y. Acad. Sci. 320: 204–213

Neal R, Gasiewicz TA, Geiger L, Olson J, Sawahata T (1984) Metabolism of 2,3,7,8-tetrachlorodibenzo-*p*-dioxin in mammalian species. In: Poland A and Kimbrough RD (eds) Banbury Report 18. Biological mechanisms of dioxin action. Cold Spring Harbor, NY: Cold Spring Harbor Laboratory. pp 49–60

Nebert DW, Gonzalez FJ (1987) P-450 genes: structure evolution and regulation. Annu. Rev. Biochem. 56: 945–993

Okey AB, Vella LM (1982) Binding of 3-methylcholanthrene and 2,3,7,8-tetrachlorodibenzo-*p*-dioxin to a common Ah receptor site in mouse and rat hepatic cytosols. Eur. J. Biochem. 127: 39–47

Okey AB, Vella LM (1984) Elevated binding of 2,3,7,8-tetrachlorodibenzo-*p*-dioxin and 3-methylcholanthrene to the Ah receptor in hepatic cytosols from phenobarbital-treated rats and mice. Biochem. Pharmacol. 33: 531–538

Okey AB, Bondy GP, Mason ME, Kahl GS, Eisen HJ, Guenthner TM, Nebert DW (1979) Regulatory gene product of the *Ah* locus: characterization of the cytosolic inducer-receptor complex and evidence for its nuclear translocation. J. Biol. Chem. 254: 11636–11648

Okey AB, Bondy GP, Mason ME, Nebert DW, Forster-Gibson CJ, Muncan J, Dufresne MJ (1980) Temperature-dependent cytosol-to-nucleus translocation of the Ah receptor for 2,3,7,8-tetrachlorodibenzo-*p*-dioxin in continuous cell culture lines. J. Biol. Chem. 255: 11415–11422

Okey AB, Mason ME, Vella LM (1983) The Ah receptor: species and tissue variation in binding of 2,3,7,8-tetrachlorodibenzo-*p*-dioxin and carcinogenic aromatic hydrocarbons. In: Rydstrom J, Montelius J and Bengtsson M (eds) Extrahepatic drug metabolism and chemical carcinogenesis. Elsevier, New York, pp. 389–393

Okey AB, Dube AW, Vella LM (1984a) Binding of benzo[a]pyrene and dibenz[a,h]anthracene to the Ah receptor in mouse and rat hepatic cytosols. Cancer Res. 44: 1426–1432

Okey AB, Vella LM, Iverson F (1984b) Ah receptor in primate liver: binding of 2,3,7,8-tetrachlorodibenzo-*p*-dioxin and carcinogenic polycyclic aromatic hydrocarbons. Can. J. Physiol. Pharmacol. 62: 1292–1295

Osborne R, Greenlee WF (1985) 2,3,7,8-Tetrachlorodibenzo-*p*-dioxin (TCDD) enhances terminal differentiation of cultured human epidermal cells. Toxicol. Appl. Pharmacol. 77: 434–443

Owens IS (1977) Genetic regulation of UDP-glucuronosyltransferase induction by polycyclic aromatic compounds in mice: co-segregation with aryl hydrocarbon (benzo(alpha)pyrene) hydroxylase induction. J. Biol. Chem. 252: 2827–2833

Pazdernik TL, Rozman KK (1985) Effect of thyroidectomy and thyroxine on 2,3,7,8-tetrachlorodibenzo-*p*-dioxin-induced immunotoxicity. Life Sci. 36: 695–703

Piskorska-Pliszczynska J, Keys B, Safe S, Newman MS (1986) The cytosolic receptor binding affinities and AHH induction potencies of 29 polynuclear aromatic hydrocarbons. Toxicol. Lett. 34: 67–74

Poellinger L, Gullberg D (1985) Characterization of the hydrophobic properties of the receptor for 2,3,7,8-tetrachlorodibenzo-*p*-dioxin. Mol. Pharmacol. 27: 271–276

Poellinger L, Lund J, Gillner M, Hansson LA, Gustafsson JA (1983) Physicochemical characterization of specific and nonspecific polyaromatic hydrocarbon binders in rat and mouse liver cytosol. J. Biol. Chem. 258: 13535–13542

Poland A, Glover E (1973) Chlorinated dibenzo-*p*-dioxins: potent inducers of delta-aminolevulinic acid synthetase and aryl hydrocarbon hydroxylase. II. A study of the structure-activity relationship. Mol. Pharmacol. 9: 736–747

Poland A, Glover E (1974) Comparison of 2,3,7,8-tetrachlorodibenzo-*p*-dioxin, a potent inducer of aryl hydrocarbon hydroxylase, with 3-methylcholanthrene. Mol. Pharmacol. 10: 349–359

Poland A, Glover E (1975) Genetic expression of aryl hydrocarbon hydroxylase by 2,3,7,8-tetrachlorodibenzo-*p*-dioxin: evidence for a receptor mutation in genetically non-responsive mice. Mol. Pharmacol. 11: 389–398

Poland A, Glover E (1979) An estimate of the maximum in vivo covalent binding of 2,3,7,8-tetrachlorodibenzo-*p*-dioxin to rat liver protein ribosomal RNA and DNA. Cancer Res. 39: 3341–3344

Poland A, Glover E (1980) 2,3,7,8-Tetrachlorodibenzo-*p*-dioxin: segregation of toxicity with the *Ah* locus. Mol. Pharmacol. 17: 86–94

Poland A, Glover E (1987) Variation in the molecular mass of the Ah receptor among vertebrate species and strains of rats. Biochem. Biophys. Res. Commun. 146: 1439–1449

Poland A, Knutson JC (1982) 2,3,7,8-Tetrachlorodibenzo-*p*-dioxin and related halogenated aromatic hydrocarbons: examination of the mechanism of toxicity. Annu. Rev. Pharmacol. Toxicol. 22: 517–554

Poland AP, Glover E, Robinson JR, Nebert DW (1974) Genetic expression of aryl hydrocarbon hydroxylase activity: induction of monooxygenase activities and cytochrome P1-450 formation by 2,3,7,8-tetrachlorodibenzo-*p*-dioxin in mice genetically "nonresponsive" to other aromatic hydrocarbons. J. Biol. Chem. 249: 5599–5606

Poland A, Glover E, Kende AS (1976) Stereospecific, high affinity binding of 2,3,7,8-tetrachlorodibenzo-*p*-dioxin by hepatic cytosol: evidence that the binding species is receptor for induction of aryl hydrocarbon hydroxylase. J. Biol. Chem. 251: 4936–4946

Poland A, Greenlee WF, Kende AS (1979) Studies on the mechanism of action of the chlorinated dibenzo-*p*-dioxins and related compounds. Ann. N.Y. Acad. Sci. 320: 214–230

Poland A, Palen D, Glover E (1982) Tumor promotion by TCDD in skin of HRS/J hairless mice. Nature (London) 300: 271–273

Poland A, Knutson JC, Glover E (1984) Histologic changes produced by 2,3,7,8-tetrachlorodibenzo-*p*-dioxin in the skin of mice carrying mutations that affect integument. J. Invest. Dermatol. 83: 454–459

Poland A, Glover E, Ebetino FH, Kende AS (1986) Photoaffinity labeling of the Ah receptor. J. Biol. Chem. 261: 6352–6365

Poland A, Glover E, Taylor BA (1987) The murine *Ah* locus: a new allele and mapping to chromosome 12. Mol. Pharmacol. 32: 471–478

Potter CL, Sipes IG, Russel DH (1983) Hypothyroxinemia and hypothermia in rats in response to 2,3,7,8-tetrachlorodibenzo-*p*-dioxin administration. Toxicol. Appl. Pharmacol. 69: 89–95

Potter CL, Moore RW, Inhorn SL, Hagen TC, Peterson RE (1986) Thyroid status and thermogenesis in rats treated with 2,3,7,8-tetrachlorodibenzo-*p*-dioxin. Toxicol. Appl. Pharmacol. 84: 45–55

Puhvel SM, Sakamoto M (1987) Responses of murine epidermal keratinocyte cultures to 2,3,7,8-tetrachlorodibenzo-*p*-dioxin (TCDD): comparison of haired and hairless genotypes. Toxicol. Appl. Pharmacol. 89: 29–36

Puhvel SM, Ertl DC, Lynberg CA (1984) Increased epidermal transglutaminase activity following 2,3,7,8-tetrachlorodibenzo-*p*-dioxin: in vivo and in vitro studies in mouse skin. Toxicol. Appl. Pharmacol. 73: 42–47

Roberts EA, Shear NH, Okey AB, Manchester DK (1985) The Ah receptor and dioxin toxicity: from rodent to human tissues. Chemosphere 14: 661–674

Romkes M, Safe S (1988) Comparative activities of 2,3,7,8-tetrachlorodibenzo-*p*-dioxin and progesterone on antiestrogens in the female rat uterus. Toxicol. Appl. Pharmacol. 92: 368–380

Romkes M, Piskorska-Pliszczynska J, Safe S (1987a) Effects of 2,3,7,8-tetrachlorodibenzo-*p*-dioxin on hepatic and uterine estrogen receptor levels in rats. Toxicol. Appl. Pharmacol. 87: 306–314

Romkes M, Piskorska-Pliszczynska J, Safe S (1987b) Quantitative structure activity relationships: analysis of interactions of 2,3,7,8-tetrachlorodibenzo-*p*-dioxin and 2-substituted analogs with rat, mouse, guinea pig and hamster cytosolic receptor. Cancer Res. 47: 5108–5111

Rozman K, Rozman T, Scheufler E, Pazdernik T, Greim H (1985) Thyroid hormones modulate the toxicity of 2,3,7,8-tetrachlorodibenzo-*p*-dioxin (TCDD). J. Toxicol. Environ. Health 16: 481–491

Ryan RP, Nelson KG, Lucier GW, Birnbaum LS, Sunihara GI (1987) 2,3,4,7,8-Pentachlorodibenzofuran and 1,2,3,4,7,8-hexachlorodibenzofuran decrease glucocorticoid receptor binding in mouse liver and placental cytosol. Toxicologist 7: 501 [Abstract]

Safe S (1984) Polychlorinated biphenyls (PCBs) and polybrominated biphenyls (PBBs): biochemistry, toxicology and mechanism of action. CRC Crit. Rev. Toxicol. 13: 319–395

Safe SH (1986) Comparative toxicology and mechanism of action of polychlorinated dibenzo-*p*-dioxins and dibenzofurans. Annu. Rev. Pharmacol. Toxicol. 26: 371–399

Safe S (1987) Determination of the 2,3,7,8-TCDD toxic equivalent factors: support for the use of the in vitro AHH induction assay. Chemosphere 16: 791–802

Safe SH (1988) The aryl hydrocarbon receptor. ISI Atlas of Science. Pharmacology 2: 78–83

Safe S, Bannister R, Davis D, Haake JM, Zacharewski T, Mayura K, Phillips TD (1988) Aroclor 1254 as a 2,3,7,8-tetrachlorodibenzo-*p*-dioxin antagonist in mice. Proceedings, Dioxin '87:

Seventh International Symposium on Chlorinated Dioxins and Related Compounds, October 4–7, Las Vegas. Available from: [University of Nevada, Las Vegas]
Sawyer T, Jones D, Rosanoff K, Mason G, Piskorska-Pliszczynska J, Safe S (1986) The biologic and toxic effects of 2,3,7,8-tetrachlorodibenzo-*p*-dioxin in chickens. Toxicology 39: 197–206
Shu HP, Paustenbach DJ, Murray FJ (1987) A critical evaluation of the use of mutagenesis, carcinogenesis, and tumor promotion data in a cancer risk assessment of 2,3,7,8-tetrachlorodibenzo-*p*-dioxin. Regul. Toxicol. Pharmacol. 7: 57–88
Sloop TC, Lucier GW (1987) Dose-dependent elevation of Ah receptor binding by TCDD in rat liver. Toxicol. Appl. Pharmacol. 88: 329–337
Sogawa K, Fujisawa-Sehara A, Yamane M, Fujii-Kuriyama Y (1986) Location of regulatory elements responsible for drug induction in the rat cytochrome P-450c gene. Proc. Natl. Acad. Sci. [USA] 83: 8044–8048
Sweatlock JA, Gasiewicz TA (1986) The interaction of 1,3-diaryltriazenes with the Ah receptor. Chemosphere 15: 1687–1690
Thomas PE, Kouri RE, Hutton JJ (1972) The genetics of aryl hydrocarbon hydroxylase induction in mice: a single gene difference between C57BL-6J and DBA-2J. Biochem. Genet. 6: 157–168
Thunberg T, Ahlborg UG, Johnson H (1979) Vitamin A (retinol) status in the rat after a single oral dose of 2,3,7,8-tetrachlorodibenzo-*p*-dioxin. Arch. Toxicol. 42: 265–274
Thunberg T, Ahlborg UG, Wahlstrom B (1984) Comparison between the effects of 2,3,7,8-tetrachlorodibenzo-*p*-dioxin and six other compounds on vitamin A storage, the UDP-glucuronosyltransferase and the aryl hydrocarbon hydroxylase activity in the rat liver. Arch. Toxicol. 55: 16–19
U.S. Environmental Protection Agency (1985) Health assessment document for polychlorinated dibenzo-*p*-dioxins. Office of Health and Environmental Assessment. 600/8-84/014F. Washington, DC. Available from: National Technical Information Service, Springfield, VA
Vecchi A, Mantovani A, Sironi M, Luini W, Spreafico F, Garattini S (1980) The effect of acute administration of 2,3,7,8-tetrachlorodibenzo-*p*-dioxin (TCDD) on humoral antibody production and cell-mediated activities in mice. Arch. Toxicol. 4: 163–165
Vecchi A, Sironi M, Canegrati MA, Recchia M, Garattini S (1983) Immunosuppressive effects of 2,3,7,8-tetrachlorodibenzo-*p*-dioxin in strains of mice with different susceptibility to induction of aryl hydrocarbon hydroxylase. Toxicol. Appl. Pharmacol. 68: 434–441
Wassom JS, Huff JE, Loprieno N (1978) A review of the genetic toxicology of chlorinated dibenzo-*p*-dioxins. Mutat. Res. 47: 141–160
Weber H, Lamb JC, Harris MW, Moore JA (1984) Teratogenicity of 2,3,7,8-tetrachlorodibenzofuran (TCDF) in mice. Toxicol. Lett. 20: 183–188
Weber H, Harris MW, Haseman JK, Birnbaum LS (1985) Teratogenic potency of TCDD, TCDF and TCDD-TCDF combinations in C57BL/6N mice. Toxicol. Lett. 26: 159–167
Whitlock JP Jr (1986) The regulation of cytochrome P-450 gene expression. Annu. Rev. Pharmacol. Toxicol. 26: 333–369
Whitlock JP Jr (1987) The regulation of gene expression by 2,3,7,8-tetrachlorodibenzo-*p*-dioxin. Pharmacol. Rev. 39: 147–161
Whitlock JP Jr, Galeazzi DR (1984) 2,3,7,8-Tetrachlorodibenzo-*p*-dioxin receptors in wild type and variant mouse hepatoma cells: nuclear location and strength of nuclear binding. J. Biol. Chem. 259: 980–985
Wilhelmsson A, Wikstrom AC, Poellinger L (1986) Polyanionic-binding properties of the receptor for 2,3,7,8-tetrachlorodibenzo-*p*-dioxin: a comparison with the glucocorticoid receptor. J. Biol. Chem. 261: 13456–13463

# 2.3 Characterization of Human Health Risks

**Risk Characterization Working Group**[1]

**C.O. Schulz, D.R. Brown and I.C. Munro**

1 Introduction . . . . . 93
2 Hazard Evaluation . . . . . 95
3 Dose-Response Characterization . . . . . 110
3.1 Low-Effect and No-Effect Levels . . . . . 110
3.2 Dose-Response Data From Carcinogenesis Bioassays . . . . . 110
4 Risk Characterization . . . . . 115
4.1 Introduction . . . . . 115
4.2 Review of Cancer Risk Assessments for 2,3,7,8-TCDD . . . . . 116
4.3 Use of the NOEL-Safety Factor Approach for Estimating a VSD for 2,3,7,8-TCDD . . . . . 119
4.4 Mechanistic Considerations in Choosing Between Alternate Approaches to Assessing Cancer Risks from Exposure to 2,3,7,8-TCDD . . . . . 119
4.5 Use of Endpoints Other Than Cancer for VSD Estimation . . . . . 121
4.6 Pharmacokinetic Considerations in the Risk Assessment for 2,3,7,8-TCDD . . . . . 121
4.7 Estimation of an ADI Based on Data From the Yusho and Yu-Cheng Incidents . . . . . 124
5 Summary and Conclusions . . . . . 129
6 References . . . . . 132

## 1 Introduction

In order to characterize the human health risks of exposure to a chemical substance or mixture of substances, it is necessary to identify the health effects that the substance(s) cause(s) in exposed humans (hazard evaluation) and then determine a quantitative relationship between the specific exposure and the incidence or severity of the effects (dose-response characterization). Ideally, information from studies of exposed human populations is used for risk characterization. In practice, in the absence of adequate data on human health effects, information from studies in experimental animals is used. This is especially true for dose-response characterization.

[1] Ad Hoc Panel on Health Aspects of Polychlorinated Dibenzo-*p*-dioxins and Polychlorinated Dibenzofurans, Universities Associated for Research and Education in Pathology, Inc., Bethesda, Maryland, USA

Environmental Toxin Series, Vol. 3

The characterization of human health risks from exposure to polychlorinated dibenzo-*p*-dioxins (PCDDs) and polychlorinated dibenzofurans (PCDFs) is made difficult by limitations in the data base available for such characterization. The data on the human health effects of PCDDs and PCDFs have been reviewed by the Epidemiology Working Group (this volume). Much of the human health effects information is derived from retrospective studies of exposed cohorts. The amounts of PCDDs and PCDFs to which members of these cohorts were exposed were not measured at the time of exposure. In most of these studies, the subjects were exposed to complex chemical mixtures. Often PCDDs and/or PCDFs were minor components of these mixtures. In addition to the lack of specific information on exposure, the interpretation of many of the epidemiologic studies of the health effects of PCDDs and PCDFs is complicated by limitations such as confounding influences, limited periods of follow-up, and the absence of measurements of biochemical indicators of possible adverse health effects. Therefore, while these studies may be useful in helping to qualitatively identify adverse health effects of PCDD/PCDF exposure in humans, they are often of limited usefulness for determining quantitative exposure-response relationships. In recent years, methods have been developed for accurately measuring concentrations of PCDDs/PCDFs in human tissues and it appears that these levels may be related to exposure in a systematic manner (Patterson et al., 1986b; Andrews et al., 1987; Kashimoto et al., 1983 as cited in Ryan et al., 1987; Kahn et al., 1988). It may be possible in future studies of exposed populations to derive quantitative estimates of exposure from measurements of total body burdens of PCDDs/PCDFs.

As shown in this and preceding chapters, 2,3,7,8-tetrachlorodibenzo-*p*-dioxin (2,3,7,8-TCDD) and some closely related PCDD and PCDF congeners are exceptionally potent and cause a wide range of toxic effects in experimental animals including cancer, impaired reproductive capability, birth defects, and suppression of immune function. In contrast, studies of human populations likely to have been exposed to PCDDs/PCDFs have not consistently shown any of these effects. The only human health effect that has been shown unequivocally to result from exposure to PCDDs/PCDFs is chloracne (see Epidemiology Working Group, this volume). The inability to link other adverse human health outcomes to PCDD/PCDF exposure could be attributable to two possible explanations. Either, humans are relatively resistant to these compounds, compared to other species, or humans have been exposed to levels of PCDDs/PCDFs that are below those that elicit measurable effects. At the present time the latter explanation appears to be the more reasonable in that limited semi-quantitative dose-response data from the Yusho and Yu-Cheng incidents suggest that humans may be as sensitive as certain experimental animals to the chloracnegenic effects of PCDFs (Ryan et al., 1987). Additional support is provided by observations that rhesus monkeys appear to be sensitive to the toxic effects of 2,3,7,8-TCDD and 2,3,7,8-tetrachlorodibenzofuran (2,3,7,8-TCDF) (Schantz et al., 1979; Bowman et al., 1987b; Hong et al., 1987).

## 2 Hazard Evaluation

Of the 210 isomers and congeners comprising the chemical classes PCDD and PCDF, 2,3,7,8-TCDD has been by far the most extensively studied in experimental animals. The effects of the closely related PCDF, 2,3,7,8-TCDF, have been studied in several species and by various routes of administration. Only a few other isomers have been studied at all and many of these studies were designed to elucidate the effects of acute or short-term exposures. Furthermore, only recently have scientists begun to investigate systematically the interactions among these compounds by testing mixtures of known composition. The available evidence indicates that the biological effects of the most toxic PCDDs and PCDFs are *qualitatively* similar (Poland and Knutson, 1982; Neal, 1985; Safe, 1986). The effects of mixtures of these compounds at submaximal doses appear to be additive (Safe, 1987). Thus, the health hazards associated with exposure to congeners other than 2,3,7,8-TCDD have been assumed to be the same as those for exposure to 2,3,7,8-TCDD, with only the dose-response relationships varying among congeners. For the purposes of this risk characterization, this assumption will be adopted unless there is direct experimental evidence to the contrary.

The biological effects which have been reported to be caused by 2,3,7,8-TCDD and structurally related compounds in experimental animals or in exposed humans are summarized in Table 1. Also shown in this table is a description of the weight of the evidence that the effect is causally related to exposure. Five classifications are used to describe the weight of evidence. "Sufficient" indicates that the effect has been demonstrated reproducibly and that the positive evidence for a cause and effect relationship strongly outweighs the evidence against such a relationship. "Limited" indicates that the effect has been shown to be associated with exposure to 2,3,7,8-TCDD but that the overall evidence for a cause and effect relationship is less than definitive. "Insufficient" describes situations in which the available evidence for a causal relationship is conflicting at the exposure level involved or is inadequate to support a positive conclusion regarding a potential association. "No evidence" indicates that no studies have been conducted that might be used either to support or deny a possible relationship. "Negative" indicates that controlled and carefully conducted studies have shown reproducibly that the effect is not causally related to exposure to 2,3,7,8-TCDD.

Separate weight-of-evidence classifications are provided in Table 1 for evidence from studies using experimental animals and for evidence from studies of exposed humans. The epidemiologic evidence that serves as the basis for the weight-of-evidence classifications for the biological effects of 2,3,7,8-TCDD in humans has been reviewed by the Epidemiology Working Group (this volume). Animal studies that serve as the basis for weight-of-evidence classifications for cancer, immunotoxicity, and reproductive toxicity are summarized in Tables 2

**Table 1.** Summary of the weight of evidence for the association of selected biological effects with exposure to 2,3,7,8-tetrachlorodibenzo-*p*-dioxin and related halogenated aromatic hydrocarbons

| Biological Effect | Weight of Evidence[1] from: Studies in animals | Studies of humans[2] |
|---|---|---|
| Alterations in serum lipid levels | Sufficient[3] | Limited |
| Cancer | Sufficient[4] | Insufficient |
| Cardiovascular toxicity | Limited[3] | Insufficient |
| Chloracne and other skin effects | Sufficient[3] (selected species) | Sufficient |
| Enzyme induction (AHH, microsomal mixed function oxidases, cytosolic enzymes) | Sufficient[5] | Sufficient[6] |
| Gastrointestinal effects (epithelial hyperplasia) | Sufficient[3] | Limited |
| Genetic toxicity | | |
| Cytogenetic toxicity | Insufficient[5] | Insufficient |
| Mutagenesis | Negative[5] | No evidence |
| Immunotoxicity | Sufficient[7] | Insufficient |
| Liver toxicity | | |
| Abnormal hepatic or urinary porphyrins | Sufficient[3] | Limited |
| Depletion of vitamin A | Sufficient[3] | No evidence |
| Organ damage | Sufficient[3] | Insufficient |
| Porphyria cutanea tarda | No animal model | Insufficient |
| Myelotoxicity | Sufficient[3] | No evidence |
| Neuro/behavioral toxicity | Insufficient[3] | Insufficient |
| Reproductive toxicity | | |
| Fetotoxicity | Sufficient[8] | Insufficient |
| Impaired reproductive capability | Sufficient[8] | Insufficient |
| Teratogenesis | Sufficient[8] (selected species) | Insufficient |
| Thymic involution | Sufficient[3] | No evidence |

[1] The criteria for the 5 different weight-of-evidence classifications are described in the text
[2] The epidemiologic studies that serve as the bases for these classifications are reviewed by the Epidemiology Working Group, (this volume)
[3] The basis for this classification is discussed in the text
[4] The basis for this classification is summarized in Tables 2 through 5
[5] The basis for this classification is summarized by the Mechanisms Working Group, (this volume)
[6] Limited evidence from human epidemiologic studies is augmented by evidence of enzyme induction in human tissues or cells in vivo (Greenlee et al., 1985a; Nagayama et al., 1985; Roberts et al., 1986; Wong et al., 1986)
[7] The basis for this classification is summarized in Table 7
[8] The basis for this classification is summarized in Table 6

through 7. The evidence regarding the ability of 2,3,7,8-TCDD to modulate enzyme activities and to cause genetic effects in experimental systems has been reviewed by the Mechanisms Working Group (this volume). The evidence that 2,3,7,8-TCDD can cause alterations in serum lipid levels in experimental animals has been reviewed by Schiller and colleagues (1986). Cardiovascular toxicity in rats and guinea pigs given single doses of 2,3,7,8-TCDD has been reported by Kelling and colleagues (1986) and Brewster and coworkers (1987). Skin lesions resembling human chloracne have been observed after dermal or

**Table 2.** Summary of studies of the carcinogenic potential of 2,3,7,8-tetrachlorodibenzo-*p*-dioxin in rodents

| Route of Adminis-tration | Species and Strain | Frequency and Duration of Dosing | Number per Group | Doses Tested | Results | Comments | Reference |
|---|---|---|---|---|---|---|---|
| Oral (feeding) | Rat/Sprague-Dawley | Continuously for 78 weeks; observed for an additional 17 weeks | 10 (males only) | 0<br>0.001<br>0.005<br>0.05<br>0.5<br>1.0<br>5.0<br>50<br>500<br>1000 ppb in diet ~ 50 pg/kg/day-50 µg/kg/day | No tumors in control or 0.001 ppb groups; 28 of 60 rats in remaining groups had tumors; no single tumor type was statistically significant; no dose-related trends | Early mortality in 3 highest dose groups; absence of tumors in controls is unusual; small group sizes | Van Miller et al., 1977 |
| Oral (feeding) | Rat/Sprague-Dawley | Continuously for 2 years | 50 per sex per group in treated groups; 83 controls | 0<br>0.022<br>0.21<br>2.2 ppb in diet (corresponds to daily doses of<br>0<br>0.001<br>0.01<br>0.1 µg/kg/day) | Statistically significant dose-related increases in the incidences of: liver tumors (females); squamous cell carcinomas of the hard palate and nasal turbinates (males & females); carcinoma of the tongue (males); carcinoma of the lung (females) | Early mortality in high dose group | Kociba et al., 1978 |
| Oral (gavage) | Mouse/Swiss/H/Riop | 1 time per week for 1 year; observed until natural death | 50 (males only) | 0<br>0.007<br>0.7<br>7.0 µg/kg/wk | Statistically significant increase in incidence of liver tumors in 0.7 µg/kg/week group, but not in high dose group | Early mortality in high dose group may have precluded increased incidence of liver tumors | Toth et al., 1978, 1979 |

**Table 2.** (*Continued*)

| Route of Administration | Species and Strain | Frequency and Duration of Dosing | Number per Group | Doses Tested | Results | Comments | Reference |
|---|---|---|---|---|---|---|---|
| Oral (gavage) | Rat/Osborne-Mendel | 2 days per week for 104 weeks | 50 per sex per group in treated groups; 75 vehicle controls per sex | 0<br>0.01<br>0.05<br>0.5 μg/kg/wk | Statistically significant dose-related increases in the incidences of: liver tumors (females); thyroid tumors (males) | Dose-related increases in subcutaneous and adrenal tumors were considered not to be conclusive by the authors | NTP, 1982a |
| Oral (gavage) | Mouse/B6C3F1 | 2 days per week for 104 weeks | 50 per sex per group in treated groups; 75 vehicle controls per sex | Males: 0<br>0.01<br>0.05<br>0.5<br>Females: 0<br>0.04<br>0.02<br>2.0<br>μg/kg/wk | Statistically significant dose-related increases in the incidences of: hepatocellular carcinoma (males and females); thyroid tumors (females); histiocytic lymphoma (females) | Increased incidences of lymphomas, leukemias, subcutaneous tumors, and lung tumors were not considered to be conclusive by the authors | NTP, 1982a |
| Dermal | Mouse/Swiss-Webster | 3 days per week for 99 weeks (males) or for 104 weeks (females) | 30 per sex per group in treated groups; 45 per sex in vehicle control group | 0<br>0.001 μg per application in males;<br>0.005 μg in females | Increased incidences of fibrosarcomas of the integument in both males and females | Volume of liquid applied was too great; high mortality, especially in males | NTP, 1982b |

**Table 3.** Summary of studies of the carcinogenic potential of 2,7-dichlorodibenzo-*p*-dioxin in rodents

| Route of Adminis-tration | Species and Strain | Frequency and Duration of Dosing | Number per Group | Doses Tested | Results | Comments | References |
|---|---|---|---|---|---|---|---|
| Oral (feeding) | Rat/Osborne-Mendel | Continuously for 110 weeks | 35 per sex per group | 0<br>5,000<br>10,000 ppm in diet | No significant differences among groups in tumor incidences at any site | — | NCI, 1979 |
| Oral (feeding) | Mouse/B6C3F1 | Continuously for 90 weeks; observed for additional 1–11 weeks | 50 per sex per group | 0<br>5,000<br>10,000 ppm in diet | Statistically significant increases in incidences of hepatocellular carcinomas in males in both treated groups | Incidence of hepato-cellular carcinomas in control was below historical range; no dose-related trend in treated groups | NCI, 1979 |

**Table 4.** Summary of studies of carcinogenic potential of a mixture of 1,2,3,6,7,8- and 1,2,3,7,8,9-hexachlorodibenzo-*p*-dioxin

| Route of Administration | Species and Strain | Frequency and Duration of Dosing | Number per Group | Doses Tested | Results | Comments | Reference |
|---|---|---|---|---|---|---|---|
| Oral (gavage) | Rat/Osborne-Mendel | 2 days per week for 104 weeks | 50 per sex per group | 0<br>1.25<br>2.5<br>5.0 µg/kg/wk | Statistically significant dose-related increases in the incidences of liver tumors in both males and females | Test substance was contaminated with 2,3,7,8-TCDD and unknown impurities; useful only for qualitative purposes | NTP, 1980a |
| Oral (gavage) | Mouse/B6C3F1 | 2 days per week for 104 weeks | 50 per sex per group | Males: 0<br>2.5<br>5.0<br>10.0<br>Females: 0<br>1.25<br>2.5<br>5.0<br>µg/kg/wk | Statistically significant dose-related increase in liver tumors in females; increased incidence of liver tumors in high dose males | Same deficiencies as the portion of the study done in rats; useful only for qualitative purposes | NTP, 1980a |
| Dermal | Mouse/Swiss-Webster | 3 days per week for 104 weeks | 30 per sex per group | 0 and 0.005 µg/mouse for 16 weeks; 0.01 µg/mouse for remainder | Increased incidence of lung cancer in males | Authors discounted findings because of comparable incidence of lung tumors in historical controls | NTP, 1980b |

**Table 5.** Summary of studies of the carcinogenic potential of 2,3,7,8-TCDD when administered with other agents

| Route of Adminis-tration | Species and Strain | Frequency and Duration of Dosing | Number per Group | Doses Tested | Results | Comments | References |
|---|---|---|---|---|---|---|---|
| Dermal | Mouse/CD-1 | Initiation/promotion; single dose of TCDD followed by TPA 2 days per week for 32 weeks | 20–30 females | 2 μg/mouse TCDD; 5 μg/mouse/application TPA | TCDD had very weak initiator activity (0.1 tumors per mouse) | TCDD was lethal to 9/30 mice | DiGiovanni et al., 1977 |
| Dermal | Mouse/CD-1 | Initiation/promotion; single dose of DMBA followed by TCDD 2 days per week of 32 weeks | 30 females | 0.1 μg/mouse/application TCDD; 200 nmol/mouse DMBA | TCDD did not increase the number of papillomas compared to DMBA alone | — | Berry et al., 1978 |
| Parenteral (i.p. injection) | Mouse/C57BL/6 or DBA/2 | Single i.p. injection of TCDD followed by single s.c. injection of 150 μg 3-MC after 48 hours | 28–98 (sex unspecified) | 0<br>1<br>100 μg/kg | TCDD did not enhance tumorigenic response to 3-MC in either strain | TCDD was lethal to 30% of C57BL/6 mice | Kouri et al., 1978 |
| Parenteral (s.c. injection) | Mouse/C57BL/6 or DBA/2 | Co-administration of TCDD and 150 μg 3-MC by s.c. injection; 36 week observation period | 28–98 (sex unspecified) | 0<br>1<br>100 μg/kg | TCDD caused a dose-related enhancement of skin tumor response to 3-MC | TCDD was lethal to many C57BL/6 mice | Kouri et al., 1978 |
| Dermal | Mouse/Sencar | Initiation/promotion; single dose of TCCD followed 72 hours later by a single dose of DMBA or B(a)P followed by TPA 2 days per week for 15 weeks | 30 females | 1 μg/mouse TCDD; 2 μg/mouse/application TPA; 10 nmol/mouse DMBA; 100 nmol/mouse B(a)P | TCDD markedly reduced the number of papillomas caused by DMBA or B(a)P followed by TPA | — | Cohen et al., 1979 |

**Table 5.** (*Continued*)

| Route of Adminis-tration | Species and Strain | Frequency and Duration of Dosing | Number per Group | Doses Tested | Results | Comments | References |
|---|---|---|---|---|---|---|---|
| Dermal | Mouse/CD-1 or Sencar | Initiation/promotion; single dose of TCDD 3 days before, 5 minutes before, or 1 day after single dose of DMBA, 3-MC, or B(a)P followed by TPA 2 days per week for 20 weeks | 30 females | 1 μg/mouse TCDD; 10 μg/mouse/ap-plication TPA; dose not available for DMBA, 3-MC, or B(a)P | TCDD markedly reduced the number of papillomas caused by DMBA, 3-MC, and B(a)P when given 3 days prior; however, TCDD slightly enhanced tumor incidences when given 1 day after DMBA, 3-MC, or B(a)P | — | DiGiovanni et al., 1979 |
| Parenteral (s.c. injection) | Rat/Charles River | Initiation/promotion; single injection of 10 mg/kg DEN to partially hepatectomized rats followed by TCDD 2 days per week for 32 weeks | 4–7 females | 0<br>0.14<br>1.4 μg/kg/injection | DEN alone and TCDD alone caused no tumors; together there was a dose-related increase in the incidence of liver tumors | — | Pitot et al., 1980 |
| Dermal | Mouse/Swiss-Webster | Initiation/promotion; single application of 50 μg DMBA followed by TCDD 3 days per week for 99 (males) or 104 (females) weeks | 30 per sex per group in treated groups; 45 controls per sex | 0<br>0.001 μg/application (females) or 0.005 μg/application (males) | No increase in tumors compared to animals treated only with TCDD | No control group initiated with DMBA and promoted with vehicle was included in the study | NTP, 1982b |

| | | | | | | | |
|---|---|---|---|---|---|---|---|
| Dermal | Mouse/HRS/J | Initiation/promotion; single application of 0.2 μmol DMBA followed by TCDD 2 days per week for 20 weeks | 20 females | 0.02 μg/ application | DMBA alone and TCDD alone caused no tumors; DMBA + TCDD caused 80% incidence of papillomas in hr/hr (hairless) mice; no effect in hr/ + (haired) phenotype | — | Poland et al., 1982, 1983 |
| Dermal | Mouse/CF-1 | Single dose of TCDD followed 24 hours later by 10 or 25 μg DMBA followed by TPA 2 days per week for 20 weeks | 20 | 0<br>0.001<br>0.01<br>0.1<br>1.0 μg/mouse | All doses of TCDD inhibited tumor formation in response to 10 μg DMBA; at 25 μg DMBA, 0.1 and 1.0 μg TCDD inhibited, but 2 lower doses enhanced tumor formation | — | Lesca, 1983 |
| Oral (feeding) | Rat/Sprague-Dawley | Initiation/promotion; partial hepatectomy and single dose of DEN followed 30 days later by feeding of diet containing TCDD and/or 2,4,5,2′,4′,5′-HCB for 20 weeks | 24 (females only) | 0<br>5 ppm HCB<br>0<br>10<br>100 ppt TCDD | TCDD or HCB alone caused no increase in enzyme altered foci (EAF) or tumors; 5 ppm HCB + 100 ppt TCDD caused increased incidence of both EAF and tumors | Abstract only | Sleight et al., 1987 |
| Oral | Rat/Unspecified | Initiation/promotion; single dose of 200 mg/kg DEN to ovariectomized female rats followed by TCDD 2 days per week for 44 or 60 weeks | N.A. | 0<br>1.4 μg/kg | TCDD caused a slight increase in enzyme altered foci at 44, but not at 60 weeks; no increase in tumor incidence compared to DEN alone | Abstract only; rats did not undergo partial hepatectomy | Goldstein et al., 1987 |

**Table 6.** Summary of studies of the reproductive toxicity/teratogenicity of TCDD in experimental animals

| Route of Administration | Species | Period of Administration | Lowest Effective Dose | Results | References |
|---|---|---|---|---|---|
| Subcutaneous injection | Mouse | Gestation | 3 μg/kg | Increased incidence of kidney anomalies | Courtney and Moore, 1971 |
| Subcutaneous injection | Rat | Gestation | 0.5 μg/kg | Increased incidence of kidney anomalies | Courtney and Moore, 1971 |
| Oral | Rat | Gestation | 0.125 μg/kg | Increased fetal death; decreased fetal weight; increased incidence of soft-tissue and skeletal abnormalities | Sparschu et al., 1971 |
| Oral | Mouse | Gestation | 3 μg/kg | Increased incidences of resorptions and cleft palate | Neubert and Dillman, 1972 |
| Oral | Mouse | Gestation | 1 μg/kg | Increased incidences of kidney anomalies and cleft palate | Moore et al., 1973 |
| Oral | Rat | Gestation | 1 μg/kg | Decreased litter size and mean fetal weight; increased incidence of soft-tissue abnormalities | Khera and Ruddick, 1973 |
| Oral | Mouse | Gestation through weaning | 2 μg/kg | Decreased spleen size and cellular immunity; suppressed graft vs host reaction | Vos and Moore, 1974 |
| Oral | Rat | Gestation through weaning | 1 μg/kg | Decreased neonatal body and thymus weight; decreased spleen size and cellular immunity | Vos and Moore, 1974 |
| Subcutaneous injection | Mouse | Gestation | 25 μg/kg | Increased incidences of kidney anomalies cleft palate, resorptions, and club foot | Courtney, 1976 |
| Oral | Mouse | Gestation | 25 μg/kg | Increased incidence of kidney anomalies | Courtney, 1976 |
| Oral | Mouse | Gestation | 1 μg/kg | Increase incidences of resorptions, cleft palate, and kidney anomalies | Smith et al., 1976 |
| Subcutaneous injection | Rat | Gestation through weaning | 5 μg/kg | Decreased spleen size and cellular immunity | Faith and Moore, 1977 |
| Subcutaneous injection | Rat | Gestation through weaning | 5 μg/kg | Decreased spleen size and cellular immunity | Faith and Luster, 1979 |

| | | | | | |
|---|---|---|---|---|---|
| Oral | Monkey | Gestation | 1 μg/kg | Increased incidence of cleft palate | Zingeser, 1979 |
| Oral (feeding) | Rat | 3-Generation reproductive study | 0.01 μg/kg/day | Decreased litter size, neonatal survival, and neonatal body weights; increased incidence of kidney anomalies | Murray et al., 1979 |
| Subcutaneous or intraperitoneal injection | Mouse | Gestation | 5–25 μg/kg | Decreased neonatal body weights; increased incidence of cleft palate | Nau and Bass, 1981 |
| Oral | Rat | Gestation | 0.5 μg/kg | Decreased neonatal body weight; increased incidence of kidney anomalies | Giavini et al., 1982a |
| Oral | Rabbit | Gestation | 0.1 μg/kg | Increased incidence of skeletal abnormalities | Giavini et al., 1982b |
| Intraperitoneal injection | Mouse | Gestation | 16 μg/kg | Increased fetal death, increased incidence of cleft palate | Hassoun and Dencker, 1982 |
| Oral | Rhesus monkey | Gestation | 1 μg/kg | Increased number of spontaneous abortions | McNulty, 1984 |
| Oral (feeding) | Rhesus monkey | 3.5–4 years | 5 ppt in diet | 25 ppt in diet caused decreased reproductive success; 5 ppt caused increased mother-infant interaction, decreased attention span, and altered learning; changes were linearly related to levels of TCDD in fat of offspring | Bowman et al., 1987a,b |

**Table 7.** Summary of studies of the effects of 2,3,7,8-tetrachlorodibenzo-*p*-dioxin on immune function in experimental animals

| Route of Administration | Species | Period of Administration | Lowest Effective Dose | Results | Reference |
|---|---|---|---|---|---|
| Oral (gavage) | Guinea pig female | Once per week for 6 weeks | 0.2 μg/kg per week | Decreased thymus weight; decreased delayed hypersensitivity to tuberculin; decreased antibody response to tetanus toxoid | Vos et al., 1973 |
| Oral (feeding) | Monkey/Rhesus female | Fed for 1 year; tested 3–4 years later | 5 ppt in diet (~0.12 ng/kg/day) | Shifts in the numbers and functions of T-lymphocytes and T-cell subsets that correlated with levels of TCDD in body fat; altered response to immunization with tetanus toxoid | Hong et al., 1987 |
| Oral (gavage) | Mouse/C57BL/6 young males | Once per week for 4 weeks | 5.0 μg/kg per week | Decreased thymus weight; decreased graft vs host response | Vos et al., 1973 |
| Oral (gavage) | Mouse/C57BL/6 | Administered to dams during gestation and postpartum to weaning | 2 μg/kg per week | Increased skin allograft rejection time | Vos and Moore, 1974 |
| Oral (gavage) | Mouse/C57BL/6 young males | Once per week for 4 weeks | 25 μg/kg per week | Decreased thymus weight; decreased lymphoproliferative response to PHA and ConA; no effect on graft vs host response | Vos and Moore, 1974 |
| Oral (gavage) | Mouse/C57BL/6 adult males | Once per week for 6 weeks | 25 μg/kg per week | Decreased thymus weight; no effect on either lymphoproliferative or graft vs host responses | Vos and Moore, 1974 |
| Oral (gavage) | Mouse/C57BL/6 males | Once per week for 4 weeks | 1 μg/kg per week | Decreased susceptibility and increased mortality due to bacterial or viral infection | Thigpen et al., 1975 |
| Oral (gavage) | Mouse/Swiss-Webster males | Once per week for 4 weeks | 1.5 μg/kg per week | Increased mortality due to E. coli endotoxin; no effect on lymphoproliferative response | Vos et al., 1978 |
| Oral (gavage) | Mouse/CD-1 males | Single dose | 10 μg/kg | No effect on thymus weight; decreased lymphoproliferative response to PHA and pokeweed; no effect on serum immunoglobulins | Sharma and Gehring, 1979 |

| | | | | | |
|---|---|---|---|---|---|
| Oral (feeding) | Mouse/Swiss-Webster | Fed to females for 4 weeks prior to mating through gestation until weaning | 1 ppb in feed 0.15 μg/kg/day | Decreased anti-SRBC response; increased mortality due to Salmonella endotoxin; decreased thymus weight; no effect on lymphoproliferative response | Thomas and Hinsdill, 1979 |
| Intraperitoneal injection | Mouse/C57BL/6 males | Once per week for 4 weeks | 0.1 μg/kg per week | Decreased number of cells in thymus and T-lymphocytes in lymph nodes; decreased anti-SRBC response; decreased generation of CTL by spleen in response to injected tumor cells | Clark et al., 1981 |
| Oral (gavage) | Mouse/C57BL/6 | Administered to dams during gestation and postpartum to weaning | 1 μg/kg per week | Decreased thymus weight; decreased bone marrow cellularity; decreased lymphoproliferative response to PHA and ConA; increased mortality due to Listeria infection; increased tumor response to injected tumor cells | Dean et al., 1981 |
| Intraperitoneal injection | Mouse/C57BL/6, DBA/2; males | Once per week for 4 weeks | 0.001 μg/kg/week in C57BL/6; 0.1 μg/kg per week in DBA/2 | Decreased generation of CTL by spleen and lymph nodes; increased mortality due to Herpes virus II | Clark et al. 1983 |
| Intraperitoneal injection or oral (gavage) | Mouse/C57BL/6; DBA/2, or C3H | Single dose (i.p.); once per week for 5–8 weeks (oral) | 1.2 μg/kg C57BL/6; 6.0 μg/kg DBA/2 | Decreased thymus weight; decrease anti-SRBC response; no effect on graft vs host response | Vecchi et al., 1983 |
| Intraperitoneal injection | Mouse/C57BL/6, DBA/2, or C3H | Once per week for 4 weeks | 0.001 μg/kg per week | Decreased generation of CTL in response to injected tumor cells | Nagarkatti et al, 1984 |
| Intraperitoneal injection | Mouse/C57BL/6 | Single injection | 1 μg/kg | No effect on thymus weight; decreased anti-SRBC response | Vecchi et al., 1985 |
| Oral (gavage) | Mouse/B6C3F1 females | Once per day for 14 days | 0.01 μg/kg/day | Decreased hemolytic complement activity; decrease in the third component of complement activity; decreased resistance to Streptococcus pneumoniae infection | White et al., 1986 |
| Oral (gavage) | Rat/CD females | Once per week for 6 weeks | 5 μg/kg per week | Decreased thymus weight; no effect on delayed hypersensitivity to tuberculin | Vos et al., 1973 |

**Table 7.** (*Continued*)

| Route of Administration | Species | Period of Administration | Lowest Effective Dose | Results | Reference |
|---|---|---|---|---|---|
| Oral (gavage) | Rat/F344 | Administered to dams during gestation and postpartum to weaning | 1 μg/kg per week | Decreased thymus weight; no effect on lymphoproliferative response to PHA; no effect on graft vs host response | Vos and Moore, 1974 |
| Oral (gavage) | Rat/F344 | Administered once per week to nursing dams | 5 μg/kg per week | Decreased thymus weight; decreased lymphoproliferative response to PHA; decreased graft vs host response | Vos and Moore, 1974 |
| Oral (gavage) | Rat/F344 | Administered to dams during gestation and postpartum to weaning | 5 μg/kg per week | Decreased thymus weight; decreased lymphoproliferative response to PHA and ConA; decreased delayed hypersensitivity response | Faith and Moore, 1977; Faith and Luster, 1979 |
| Oral (gavage) | Rat/F344 | Administered once per week to nursing dams | 5 μg/kg per week | Decreased thymus weight; decreased lymphoproliferative response to PHA and ConA; decreased delayed hypersensitivity | Luster et al., 1982 |

parenteral administration of 2,3,7,8-TCDD to monkeys (McConnell et al., 1978), certain strains of mice (Poland et al., 1982), dairy cattle (McConnell, 1985), and rabbits (Jones and Krizek, 1962).

The reported effects of 2,3,7,8-TCDD on the gastrointestinal and urinary tracts of experimental animals include hyperplasia of the mucous-producing cells in the stomach (McNulty, 1977; McConnell et al., 1978) and proliferative changes in the intestines and bile ducts and in the transitional epithelium lining the urinary tract (McConnell, 1985). The effects of 2,3,7,8-TCDD on the liver have been reviewed by Sweeney and coworkers (1984). Relatively high doses cause vacuolization, lipid accumulation, proliferation of the rough and smooth endoplasmic reticulum, and cellular necrosis (Weber et al., 1983). A relatively more sensitive indicator of effects on the liver is an altered porphyrin excretion pattern (Kociba et al., 1976). Large single doses of 2,3,7,8-TCDD also deplete stores of vitamin A in the liver (Thunberg, 1984).

Adverse effects of 2,3,7,8-TCDD on the bone marrow (myelotoxicity) have been demonstrated in monkeys (Allen et al., 1977; McNulty, 1977; McNulty et al., 1981), rats (Kociba et al., 1976), and mice (Luster et al., 1985). The neurobehavioral toxicity of 2,3,7,8-TCDD has not been systematically studied in experimental animals although subtle behavioral effects have been reported in perinatally exposed monkeys (Bowman et al., 1987a). Finally, thymic involution is regularly seen in animals exposed to doses of 2,3,7,8-TCDD that result in acute toxicity (McConnell, 1985).

Ultimately, for the purposes of risk characterization, it is necessary to evaluate the weight of evidence from studies of both experimental animals and exposed humans in order to identify those biological effects which are most likely to manifest themselves as adverse health effects in humans exposed to low levels of PCDDs and PCDFs in the environment. Some of the biological effects described in Table 1 need not be considered in assessing the potential human health impact of long-term exposure to PCDDs/PCDFs as the result of exposure to relatively low environmental concentrations. Effects such as cardiovascular toxicity, porphyria cutanea tarda, genetic toxicity, and neurobehavioral toxicity can be eliminated as the bases for estimating human health risks, because the available data base does not support a conclusion regarding causality in either humans or animals.

Alterations in serum lipid levels, enzyme induction in the liver and other tissues, and alterations in hepatic and urinary porphyrin patterns cannot necessarily be considered to be adverse health effects of and by themselves. It is not possible to relate these biological indicators quantitatively to possible toxicological endpoints without further study.

The available dose-response information from studies in experimental animals indicates that several of the effects shown in Table 1 occur only as a result of short-term exposure to doses of 2,3,7,8-TCDD that are higher than those required to induce other manifestations of toxicity. These include chloracne and other skin effects (e.g. epidermal hyperplasia), myelotoxicity, gastrointestinal hyperplasia, liver damage, depletion of vitamin A, and thymic involution.

The toxic endpoints that are left for consideration as possible bases for the characterization of human health risks attributable to long-term exposure to low levels of PCDDs/PCDFs in the environment are cancer, immunotoxicity, and reproductive toxicity, i.e., impaired reproductive capacity, fetotoxicity, teratogenicity, and developmental toxicity. Each of these endpoints has been conclusively shown to be caused by exposure to 2,3,7,8-TCDD and related compounds in experimental animals. None of these effects have been conclusively demonstrated in humans who may have been exposed to these compounds. However, the available data from studies of exposed humans is inadequate to support a conclusion that such effects would not occur at doses comparable to those given to experimental animals.

## 3 Dose-Response Characterization

### *3.1 Low-Effect and No-Effect Levels*

Table 8 summarizes information from experiments in animals on low effect levels (LELs) of 2,3,7,8-TCDD of some indicators of biological activity. It is important to recognize that in each case the LEL caused a statistically and/or biologically significant effect in the species tested. In a few instances a lower dose was tested and met the criteria for a no-effect-level (NOEL). In most cases, however, the LEL was the lowest dose tested. The data in this table support several general conclusions. First, the cumulative toxicity of 2,3,7,8-TCDD is very apparent from the decrease in LEL (expressed as a daily dose) with the duration of exposure. As a rough approximation, chronic LELs expressed as daily doses are smaller than acute single dose LELs for similar effects by a factor of approximately $10^3$. Second, for several effects, monkeys appear to be at least as sensitive as any other species tested. Finally, the LELs for a number of different effects appear to be within one or two orders of magnitude of each other. This observation provides support for considering endpoints other than cancer in the characterization of human health risks from exposure to PCDDs and PCDFs and may also imply some mechanistic commonality in the causation of these effects.

### *3.2 Dose-Response Data From Carcinogenesis Bioassays*

Because of the absence of consistent data indicating an increase in cancer in studies of exposed humans, it is necessary to turn to the results of studies in experimental animals in order to obtain dose-response data. Only two studies have shown statistically significant increases in the incidences of single tumor

**Table 8.** Low effect levels for the toxic effects of 2,3,7,8-tetrachlorodibenzo-*p*-dioxin in experimental animals

| Effect | Species and Strain | Route | Dose and Duration | Reference |
|---|---|---|---|---|
| Acute lethality | Guinea pig | Oral | 2.0 μg/kg single dose | McConnell et al., 1978 |
| Cancer | Sprague-Dawley rat | Oral (feeding) | 210 ppt in diet for 2 years (10 ng/kg/day) | Kociba et al., 1978 |
| Cancer promotion | HRS/J hairless mouse | Dermal | 7.5 ng/week for 20 weeks | Poland et al., 1982, 1983 |
| | Charles River rat | Injection (s.c) | 280 ng/kg/week for 32 weeks | Pitot et al., 1980 |
| Chloracne | Rabbit | Dermal to ear | 4 ng/day, 5 days/week for 4 weeks (0.4 ng/day = no effect level) | Schwetz et al., 1973 |
| Chronic lethality | Rhesus monkey | Oral (feeding) | 500 ppt in diet for 9 months ( ~ 12 ng/kg/day) | Allen et al., 1977 |
| Chronic toxicity (hair loss, hyperkeratosis, body weight loss) | Rhesus monkey | Oral (feeding) | 50 ppt in diet for 20 months ( ~ 1.2 ng/kg/day) | Schantz et al., 1979 |
| Hepatic effects: | | | | |
| Altered porphyrin metabolism | Sprague-Dawley rat | Oral (feeding) | 210 ppt in diet for 2 years (10 ng/kg/day) | Kociba et al., 1978 |
| Histopathological changes | Sprague-Dawley rat | Oral (feeding) | 21 ppt in diet for 2 years (1 ng/kg/day) | Kociba et al., 1978 |
| Increased lipid peroxidation | Sprague-Dawley rat | Oral | 1 μg/kg single dose | Stohs et al., 1986 |
| Lipid accumulation | Sprague-Dawley rat | Oral (feeding) | 100 ng/kg/day for 13 weeks | Kociba et al., 1976 |
| Immunotoxicity: | | | | |
| Alterations in T-cell subsets | Rhesus monkey | Oral (feeding) | 5 ppt in diet for 1 year ( ~ 0.12 ng/kg/day) | Hong et al., 1987 |
| Suppression of CTL generation | C57BL/6 mouse | Injection (i.p.) | 1 ng/kg/day, 1 day/week for 4 weeks | Clark et al., 1983; Nagarkatti et al., 1984 |
| Suppression of serum complement activity | B6C3F1 mouse | Oral | 10 ng/kg/day for 14 days | White et al., 1986 |

**Table 8.** (*Continued*)

| Effect | Species and Strain | Route | Dose and Duration | Reference |
|---|---|---|---|---|
| Induction of AHH | Sprague-Dawley rat | Oral | 2.0 ng/kg single oral dose (0.6 ng/kg = no effect level) | Kitchin and Woods, 1978 |
| | Sprague-Dawley rat | Oral | 10 ng/kg/day, 2 days/week for 22 weeks | Sloop and Lucier, 1987 |
| Reproductive effects: | | | | |
| Impaired conception & spontaneous absorptions | Sprague-Dawley rat | Oral (feeding) | 10 ng/kg/day in the diet for 1 year | Murray et al., 1979 |
| | Rhesus monkey | Oral (feeding) | 5 ppt in diet for 6 months to 4 years prior to mating ( ~ 0.12 ng/kg/day; estimated no effect level = 0.05 ng/kg/day) | Bowman et al., 1987b |
| Perinatal toxicity: | | | | |
| Decreased thymus weight | Wistar rat | Oral | 250 ng/kg to dams on day 16 of gestation | Madsen and Larsen, 1987 |
| Induced hepatic AHH and ECOD | Wistar rat | Oral | 100 ng/kg to dams on day 16 of gestation | Madsen and Larsen, 1987 |
| Teratogenesis: | | | | |
| Cleft palate | CF-1 mouse | Oral | 1 μg/kg to dams on days 6–15 of gestation | Smith et al., 1976 |
| | Rhesus monkey | Oral | 1 μg/kg/day to dams on days 20–40 of gestation | Zingeser, 1979 |
| Kidney anomalies | Sprague-Dawley rat | Injection (s.c.) | 0.5 μg/kg/day to dams on days 6–15 of gestation | Courtney and Moore, 1971 |
| | C57BL/6 mouse | Oral | 1 μg/kg/day to dams on days 10–13 of gestation | Moore et al., 1973 |
| | Sprague-Dawley rat | Oral (feeding) | 10 ng/kg/day for 1 year prior to mating | Murray et al., 1979 |

types in animals treated with 2,3,7,8-TCDD. These are the studies of Kociba et al. (1978) and the National Toxicology Program (NTP) (1982a). The incidences of those tumors that appear to be increased as a result of exposure to the test substance are shown in Tables 9 and 10. It can be seen from these tables that the incidences of many of the tumors increased in a dose-dependent fashion. These include: hepatocellular tumors in female rats and squamous cell carcinomas of the tongue, hard palate, or nasal turbinates in male and female rats in the Kociba study; and hepatocellular tumors in male and female rats and male and female mice, thyroid tumors in male rats and female mice, and subcutaneous fibromas or fibrosarcomas in male and female rats and female mice in the NTP bioassays. Two pathologists have evaluated the liver lesions in animals in the Kociba study and have arrived at slightly different tumor incidences. The appearance of both benign and malignant tumors in the same tissue (liver) in animals in these studies raises the scientific issue of the appropriateness of combining their incidences for the purpose of estimating dose-response relationships. An additional issue is the increased incidences of primary tumors at multiple sites in these animals.

Examination of the tumor incidence data in Tables 9 and 10 indicates that the liver is an important target for the tumorigenic activity of 2,3,7,8-TCDD in rodents. Statistically significant excesses of liver tumors were seen in female rats in the Kociba study and in both male and female rats and male and female mice in the NTP study. It has been argued that enhancement of liver tumor incidence

**Table 9.** Percentages (%) of animals having tumors at selected sites in the Kociba et al. (1978) bioassay of 2,3,7,8-tetrachlorodibenzo-*p*-dioxin

| Tumor | Pathologist[1] | Daily Doses (μg/kg) 0 | 0.001 | 0.01 | 0.1 |
|---|---|---|---|---|---|
| Male Rats | | | | | |
| Stratified squamous cell carcinoma | Kociba | 0 | 0 | 0 | 14 |
| of hard palate or nasal turbinates | Squire | 0 | 3 | 0 | 20 |
| Squamous cell carcinoma of the | Kociba | 0 | 2 | 2 | 7 |
| tongue | Squire | 0 | 4 | 2 | 7 |
| Total combined tumors | Squire | 0 | 4 | 2 | 20 |
| Female Rats | | | | | |
| Hepatocellular neoplastic nodule | Kociba | 9 | 6 | 36 | 47 |
| Hepatocellular carcinoma | Kociba | 1 | 0 | 4 | 22 |
| Hepatocellular neoplastic nodule | Kociba | 10 | 6 | 36 | 68 |
| or carcinoma | Squire | 19 | 16 | 54 | 70 |
| Stratified squamous cell carcinoma | Kociba | 0 | 0 | 4 | 17 |
| of the hard palate or nasal turbinates | Squire | 0 | 0 | 4 | 23 |
| Keratinizing squamous cell | Kociba | 0 | 0 | 0 | 14 |
| carcinoma of the lung | Squire | 0 | 0 | 0 | 17 |
| Total combined tumors | Kociba | 10 | 6 | 36 | 69 |
| | Squire | 19 | 16 | 54 | 72 |

[1] Incidence data designated as Kociba was that reported in the published version of the bioassay. At a later date, tissue slides were independently evaluated by another pathologist, R. Squire, (U.S. Environmental Protection Agency, 1985), to yield slightly different incidence data

**Table 10.** Percentages of animals with tumors at selected sites in the National Toxicology Program (1982a) bioassays of 2,3,7,8-tetrachlorodibenzo-*p*-dioxin

| Tumor | Weekly Doses (μg/kg) | | | |
|---|---|---|---|---|
| | 0 | 0.01 | 0.05 | 0.5 |
| Male Rats | | | | |
| Thyroid follicular cell adenoma | 0 | 10 | 12 | 20 |
| Thyroid follicular cell adenoma or carcinoma | 1 | 10 | 16 | 22 |
| Hepatocellular carcinoma | 0 | 0 | 0 | 2 |
| Hepatocellular neoplastic nodule, adenoma, or carcinoma | 0 | 0 | 0 | 8 |
| Subcutaneous fibroma | 4 | 2 | 6 | 14 |
| Female Rats | | | | |
| Hepatocellular neoplastic nodule | 7 | 2 | 6 | 24 |
| Hepatocellular carcinoma | 0 | 0 | 0 | 4 |
| Hepatocellular neoplastic nodule or carcinoma | 7 | 2 | 6 | 28 |
| Subcutaneous fibrosarcoma | 0 | 4 | 6 | 8 |
| Male Mice | | | | |
| Hepatocellular adenoma | 10 | 6 | 10 | 20 |
| Hepatocellular carcinoma | 11 | 18 | 16 | 34 |
| Hepatocellular adenoma or carcinoma | 20 | 24 | 26 | 54 |
| Female Mice | | | | |
| Hepatocellular carcinoma | 1 | 4 | 4 | 13 |
| Hepatocellular adenoma or carcinoma | 4 | 12 | 12 | 23 |
| Thyroid follicular cell adenoma | 0 | 6 | 2 | 11 |
| Lymphoma or leukemia | 24 | 24 | 27 | 45 |
| Subcutaneous fibrosarcoma | 1 | 2 | 2 | 11 |

in rodents has little relevance for human cancer risk characterization because these tumors occur spontaneously; whereas, the incidence of liver tumors in humans is low (Society of Toxicology, 1982). However, in the case of 2,3,7,8-TCDD a significant excess of liver tumors was seen in male rats in the NTP study where there was no spontaneous background incidence in the untreated controls. The other statistically significant excess incidences of cancer seen in the Kociba study were cancers of the hard palate, tongue, nasal turbinates, and lungs. None of these sites showed excess incidences of cancer in the NTP study.

Significant increases in the incidences of tumors of the thyroid were seen in male rats and female mice in the NTP bioassays but were not seen in the rats in the Kociba study. These tumors are relatively rare and not thought to occur spontaneously in the animal strains used in these studies (Altman and Goodman, 1979). Also, it is the only tumor for which the incidence was significantly elevated at the lowest dose tested (male rats in the NTP study).

Finally, administration of 2,3,7,8-TCDD resulted in significantly increased incidences of subcutaneous fibromas in male rats and subcutaneous fibrosarcomas in female rats and mice in the NTP studies. In rats, these tumors are relatively common spontaneous tumors with incidences of fibromas ranging from 1–22% and for fibrosarcomas from 1–74% in control animals (Altman and Goodman, 1979). The incidences of these tumors in the NTP bioassays are well within these ranges and are therefore probably of no biological significance.

## 4 Risk Characterization

### *4.1 Introduction*

Several approaches have been used for the quantitative estimation of "virtually safe doses" (VSDs) or "acceptable daily intakes" (ADIs) for exposure to toxic chemicals (National Research Council, 1980). One general approach is to determine a probable NOEL from studies of exposed human populations or from studies in experimental animals. The ADI is then calculated by dividing the NOEL by an appropriate safety factor to account for uncertainties in the experimentally determined NOEL. Another general approach is to fit a mathematical model to the observed dose-response data and to use the model to predict responses at doses other than those to which the test animals or study population were exposed. The VSD is estimated by using the mathematical model to calculate the dose at which the response is either zero or (for non-threshold models) some arbitrarily low level. For regulatory purposes the virtually safe dose for cancer corresponds to a lifetime risk of $10^{-5}$ or $10^{-6}$ (U.S. Office of Science and Technology Policy, 1985). In order to use the mathematical approach, response data must be expressed in quantitative terms such as percent incidence or some quantitative measure of severity. Furthermore, there must be a minimum of two dose levels at which non-zero responses are observed in order for the goodness of fit of the mathematical model to be tested. The NOEL-safety factor approach is much less demanding in its data requirements and can be applied to data from studies in which there was only one non-zero exposure level so long as that level was a NOEL.

Traditionally, regulatory agencies in the United States have used the mathematical model approach to estimate VSDs for exposure to chemicals that are presumed to cause human cancer, whereas the NOEL-safety factor approach has been used to estimate ADIs (also referred to as reference doses) for compounds that cause toxic effects other than cancer (National Research Council, 1980; Krewski et al., 1984). The rationale for this divergence in approach is that cancer is a self-replicating toxic effect (Saffiotti, 1977). In other words, the interaction between an ultimate carcinogen and the genetic material of a cell may initiate a chain of events leading to malignancy, even in the absence of additional exposure. Under this hypothesis any non-zero exposure to a carcinogen results in a finite probability of a response. In contrast, the traditional regulatory approach has been to assume that toxic effects other than cancer are not self-replicating. In other words, there is some non-zero level of exposure at which the interaction betweeen the compound and the cell results in no grossly detectable damage and neither the severity nor incidence of the effect increases in the absence of additional exposure. For such effects, this traditional approach assumes the presence of no adverse effect levels.

The traditional distinction between cancer and other toxic effects on the basis of theoretical mechanisms of action is probably simplistic. It is likely that toxic effects other than cancer, e.g. autoimmune disorders and heritable genetic defects, are self-replicating. On the other hand, it is clear that a number of compounds that cause increased incidences of cancer in animals do so via mechanisms that may not involve self-replicating damage to the genetic material of cells (Butterworth and Slaga, 1987). Such compounds have come to be referred to as non-genotoxic carcinogens. The hypothetical mechanisms that have been proposed for some non-genotoxic carcinogens suggest that the carcinogenic actions of these agents are reversible and that the concept of a finite probability of an effect at any non-zero dose probably does not apply (Pitot et al., 1981). Thus, the use of non-threshold mathematical models to estimate dose-response relationships may not be appropriate for certain classes of non-genotoxic carcinogens.

## *4.2 Review of Cancer Risk Assessments for 2,3,7,8-TCDD*

The choice of tumor endpoint has considerable impact on the estimated VSD if the mathematical modeling approach is used for risk characterization. Table 11 shows the VSDs of 2,3,7,8-TCDD that are calculated using a linearized multi-

**Table 11.** Virtually safe doses (VSDs)[a] for 2,3,7,8-tetrachlorodibenzo-*p*-dioxin calculated using the linearized multistage model with different tumor incidence data[b]

| Tumor | Species | Sex | Study | Pathol. | VSD |
|---|---|---|---|---|---|
| Hepatocellular adenoma or carcinoma | Rat | M | NTP[c] | | 2,600 |
| Hepatocellular carcinoma | Rat | M | NTP | | 1,600 |
| Hepatocellular carcinoma | Rat | F | NTP | | 1,300 |
| Hepatocellular adenoma or carcinoma | Rat | F | NTP | | 130 |
| Hepatocellular carcinoma | Mouse | F | NTP | | 2.5 |
| Hepatocellular adenoma or carcinoma | Mouse | F | NTP | | 1.4 |
| Carcinoma of the tongue | Rat | M | Kociba[d] et al. | Kociba | 1.5 |
| | | | | Squire | 1.4 |
| Carcinoma of the hard palate or nasal turbinates | Rat | M | Kociba et al. | Kociba | 0.78 |
| | | | | Squire | 0.49 |
| Carcinoma of the hard palate or nasal turbinates | Rat | F | Kociba et al. | Kociba | 0.50 |
| | | | | Squire | 0.37 |
| Carcinoma of the lung | Rat | F | Kociba et al. | Kociba | 0.73 |
| | | | | Squire | 0.60 |
| Hepatocellular carcinoma | Mouse | M | NTP | | 0.26 |
| Hepatocellular adenoma or carcinoma | Mouse | M | NTP | | 0.13 |
| Hepatocellular carcinoma | Rat | F | Kociba et al. | Kociba | 0.40 |
| Hepatocellular adenoma or carcinoma | Rat | F | Kociba et al. | Kociba | 0.077 |
| | | | | Squire | 0.086 |

[a] VSD = the daily dose in picograms per kilogram of body weight corresponding to an added cancer risk of one in one million; [b] The data in this table was compiled from Portier et al. (1984) and from Sielken (1987); [c] National Toxicology Program (1982a); [d] Kociba et al., 1978

stage mathematical model and various tumor incidence data from two different carcinogenesis bioassays. There is a 30,000-fold range of estimated VSDs depending on the dose-response data selected. Interestingly, the differences between pathologists with regard to the incidences of liver tumors in the Kociba et al. (1978) study have only a small effect on estimated VSDs. Also, the combining of benign and malignant liver tumors has a small effect on the estimated VSD, generally lowering the VSD by a factor of ten or less. However, utilizing the hepatocellular tumors in the male rats in the NTP study for mathematical modeling actually increases the estimate of the VSD (National Toxicology Program, 1982a). The VSDs for combined hepatocellular tumors in female rats calculated using data from the two different bioassays were different by more than three orders of magnitude. There is no scientific basis for selecting any one tumor endpoint in any given species or sex as the endpoint that is most predictive for human health risk assessment. Traditionally, the U.S. Environmental Protection Agency (EPA) has taken the approach that "the biologically acceptable data set from long-term animal studies showing the greatest sensitivity should generally be given the greatest emphasis" (U.S. Environmental Protection Agency, 1984). This approach reflects a policy of the regulatory agencies to err on the side of overestimation of risk.

Another major contributor to uncertainty in the estimation of cancer risks due to 2,3,7,8-TCDD using the mathematical modeling approach is the selection of the model. The U.S. Environmental Protection Agency (1985) and the Centers for Disease Control (CDC) (Kimbrough et al., 1984) have selected the linearized multistage model. Historically, the regulatory agencies have justified the selection of this model for dose-response relationships on the basis that it is derived from theoretical considerations regarding the mechanisms by which chemicals induce cancer. Also, of the mathematical models commonly used to fit experimental dose-response data, the linearized multistage model is generally more conservative at doses below the observed range, i.e., at any given dose it estimates a higher probable response than do the other models. This is the result of the linear nature of the model in the low-dose range, whereas other models (e.g. Weibull and probit) are sublinear in this range. This generalization does not hold for 2,3,7,8-TCDD. The U.S. Environmental Protection Agency (1985) compared human cancer risk estimates which were derived using four different mathematical models applied to the dose-response data for total tumor incidence in female rats in the study by Kociba and colleagues (1978). The VSD predicted by the one-hit model was identical to that predicted by the linearized multistage model, whereas the VSD predicted by the Weibull model was approximately two orders of magnitude *lower* than that predicted by the linearized multistage model, apparently due to the high curvature of the dose-response curve in the observed range. The VSD predicted by the log-probit model was approximately two orders of magnitude higher than that predicted by the linearized multistage model.

Table 12 presents VSDs for 2,3,7,8-TCDD estimated using mathematical models that predict a finite risk at every non-zero dose and compares them to

**Table 12.** Representative estimated virtually safe doses (VSDs) or acceptable daily intakes (ADI) for 2,3,7,8-tetrachlorodibenzo-*p*-dioxin

| Source | Approach | Data Base Used | VSD or ADI |
|---|---|---|---|
| U.S. EPA, 1985 | Linearized multistage model | All tumors in female rats (Kociba et al., 1978) | 6.4 fg/kg/day |
| U.S. CDC[a] (Kimbrough et al., 1984) | Linearized multistage model | Hepatocellular tumors in female rats (Kociba et al., 1978) | 28 fg/kg/day |
| Netherlands (van der Heijden et al., 1982) | NOEL-Safety factor | Hepatocellular tumors in female rats (Kociba et al., 1978) | 4 pg/kg/day |
| Ontario, 1985 | NOEL-Safety factor | Tumors in rats (Kociba et al., 1978) | 10 pg/kg/day |
| U.S. FDA[b] (Cordle, 1981) | NOEL-Safety factor | Tumors in rats (Kociba et al., 1978) | 13 pg/kg/day |
| Sielken, 1987 | Time-to-tumor and multistage models fitted to 2 lower dose groups | Hepatocellular tumors in female rats (Kociba et al., 1978) | 100–5,000 pg/kg/day |

[a] U.S. Centers for Disease Control; [b] U.S. Food and Drug Administration

ADIs calculated using the NOEL and safety factor approach discussed below. The VSD estimated by the EPA differs from those estimated by Kimbrough and coworkers (1984) for several reasons. First, EPA used incidence data for all tumor sites combined from the Kociba et al. (1978) study. Second, EPA calculated the human equivalent dose using the dose per unit of body surface area. Finally, CDC used final concentrations of 2,3,7,8-TCDD in the livers of treated rats as a surrogate for dose and calculated equivalent doses in humans using relative body weights.

Sielken (1987) has strongly criticized the use of the linearized multistage model for estimating dose response relationships for 2,3,7,8-TCDD below the observed range. His objection is based on the poorness of fit of the model to the data in the observed range and the sensitivity of the model to the incidence data from the highest dose tested. Sielken argues that better and more consistent estimates of the dose-response relationship in the low dose range can be achieved by using either a time-to-tumor model or a multistage model fitted to the response seen at the two lower doses in the Kociba et al. (1978) study. Using these approaches and the data on the incidence of liver tumors in female rats from the Kociba et al. (1978) study, he estimated that the VSD for rats is in the range of from 0.1 to 5 ng/kg/day. This is three to six orders of magnitude higher than VSDs estimated using the linearized multistage model applied to all the dose response data. Although Sielken does not provide a clear, biologically-based rationale to justify either the weighting of tumor responses at the lower dose levels or the use of a time-to-tumor model, his arguments, nevertheless, provide useful insights on the sensitivity of risk estimates for 2,3,7,8-TCDD to such variables as model selection and the choice of dose-response data.

### *4.3 Use of the NOEL-Safety Factor Approach for Estimating a VSD for 2,3,7,8-TCDD*

As shown in Table 12 several regulatory authorities, primarily outside of the United States, have adopted the NOEL-safety factor approach to the estimation of an ADI for the carcinogenic effect of 2,3,7,8-TCDD (van der Heijden et al., 1982; Ontario, 1985). These approaches are similar and take the lowest dose level (1 ng/kg/day) from the Kociba et al. (1978) study as the NOEL. The Ontario Ministry of Health used a safety factor of 100 to calculate an ADI of 10 pg/kg/day, whereas the Government of the Netherlands used a safety factor of 250. The differences of the three ADIs estimated using the NOEL-safety factor approach are attributable principally to the magnitude of the safety factors selected.

### *4.4 Mechanistic Considerations in Choosing Between Alternate Approaches to Assessing Cancer Risks from Exposure to 2,3,7,8-TCDD*

Proponents of the NOEL-safety factor approach to deriving a safe level for human exposure to 2,3,7,8-TCDD argue that the use of mathematical models that predict a finite cancer risk at every non-zero exposure level is inappropriate because of mechanistic considerations (Shu et al., 1987). According to this argument, 2,3,7,8-TCDD is incapable of initiating the carcinogenic process in normal cells. Rather, the increased incidence of cancer in experimental animals treated with 2,3,7,8-TCDD is the result of promotion of already initiated cancer cells in the animal. This promotional effect is reversible and non-self-replicating. Based on these assumptions, these authors conclude that tumor promotion is similar to other non-carcinogenic toxic effects and a NOEL-safety factor approach is appropriate for estimating an ADI.

The evidence supporting the above argument is largely negative in nature. In other words, 2,3,7,8-TCDD does not cause gene mutations in genetic toxicity test systems and it does not behave as an initiator in classical initiation/promotion assays (see Mechanisms Working Group, this volume; and Table 5). The positive evidence consists of the tumor promoting activity seen in the short term in vivo assays by Pitot and coworkers (1980) and Poland and coworkers (1982), and the ability of 2,3,7,8-TCDD to cause morphologic transformation and alter intercellular communication in cultured mouse fibroblasts (Abernethy et al., 1985; Boreiko et al., 1986, 1987; Abernethy and Boreiko, 1987). On the other hand, 2,3,7,8-TCDD was relatively inactive as a promoter in assays reported by Berry and coworkers (1978) and Goldstein and coworkers (1987). These latter studies are not precisely comparable to the positive studies because of methodological differences.

The weakness in the above argument is that the distinction between initiators and promoters is an operational one based on theoretical considerations and the behavior of compounds in test systems. The mechanisms by which chemicals in general, initiate and modify the carcinogenic process are not known. The mechanism by which 2,3,7,8-TCDD increases cancer incidence in experimental animals has not been systematically investigated. Bioassays involving interrupted dosing schedules and/or serial sacrifices have not been performed, nor have promotional assays been conducted in such a way as to assess the reversibility of the promotional process.

Some information on the mechanism by which 2,3,7,8-TCDD promotes cancer in experimental models is available. Based on the segregation of the promotional effect with specific genetic loci in mice, Poland (1984) suggests that binding of 2,3,7,8-TCDD to the Ah receptor is a necessary but insufficient step in the process. He also suggests that the *hr* gene locus is involved in this mechanism. According to this hypothesis, promotion of tumorigenesis is simply another component of the pleiotropic response elicited by the interaction of the 2,3,7,8-TCDD-Ah receptor complex with the genome and may be the result of the activation of structural and regulatory genes. Whereas the binding of 2,3,7,8-TCDD to the Ah receptor is a reversible process, the reversibility of subsequent steps in this process is unknown.

Additional mechanistic evidence is provided by a number of studies of mammalian cells in culture which indicate that 2,3,7,8-TCDD and structurally related compounds stimulate cell proliferation and terminal differentiation (Knutson and Poland, 1984a,b; Rice and Cline, 1984; Greenlee et al., 1985a,b; Puhvel et al., 1985; Puhvel and Sakamoto, 1987). Citing these studies, some authors (Shu et al., 1987) have suggested that 2,3,7,8-TCDD may increase the incidence of cancer by enhancing the proliferation of initiated cells.

Another possible mechanism by which 2,3,7,8-TCDD may increase cancer incidence in experimental animals is through modulation of the immune system. It has been shown to decrease bone marrow cellularity, inhibit the maturation of T-cells in the thymus, and interfere with the generation of cytotoxic T lymphocytes in response to injected tumor cells in mice (Dean et al., 1981; Clark et al., 1981, 1983). Cytotoxic T lymphocytes have been postulated as being important agents in cancer immunosurveillance (Dean et al., 1986). Impaired immunosurveillance is also consistent with the observation that 2,3,7,8-TCDD may increase the incidence of cancer at a variety of different sites.

In summary, the available information on the mechanism by which 2,3,7,8-TCDD increases the incidence of cancer in experimental animals is preliminary and complex. The evidence that it is not genotoxic and that it promotes tumor development in otherwise initiated tissues is quite convincing and supports the hypothesis that the early stages of its carcinogenic actions are potentially reversible and not self-replicating. On the other hand, it is not possible to rule out irreversible steps and self-replicating damage in the overall process by which the carcinogenic response is enhanced. It is not possible to choose between currently accepted approaches to risk assessment on the basis of mechanistic considerations alone.

### *4.5 Use of Endpoints Other Than Cancer for VSD Estimation*

Those estimates of an ADI for 2,3,7,8-TCDD using the NOEL and safety factor approach are subject to the criticism that they use a limited data base for the selection of the NOEL. In general, all these approaches choose the lowest dose level tested (1 ng/kg/day) in the Kociba et al. (1978) bioassay as a NOEL. The Ontario (1985) group also cited the three-generation reproduction study of Murray and coworkers (1979) as supporting a NOEL of 1 ng/kg/day. The conclusion from both studies, that this value represents a NOEL, may be questionable. There was no NOEL for the increased incidence of thyroid follicular cell adenomas in male rats in the NTP (1982a) bioassays. The lowest dose tested corresponded to an average daily dose of 1.4 ng/kg/day which strongly suggests that 1 ng/kg/day is not a NOEL for this tumor in rats. The conclusion of Murray and coworkers (1979) that 1 ng/kg/day was a NOEL in their study has been challenged by Nisbet and Paxton (1982) who argued that a more appropriate statistical analysis of the data indicated that there were decrements in reproductive performance at the lowest dose tested in this study.

If a NOEL-safety factor approach is to be used in estimating a VSD for 2,3,7,8-TCDD based on the argument that the promotion of cancer by this compound is reversible and exhibits a threshold, then there is no rational basis for considering only cancer in establishing a NOEL. A number of effects are seen at doses comparable to those that produce cancer (see Table 8). Very few of these studies have demonstrated clear NOELs. This is particularly true of long-term chronic studies in rodents and rhesus monkeys. Sensitive indicators of immune function have not been evaluated in any of the available lifetime bioassays of 2,3,7,8-TCDD in rodents. The available evidence suggests that NOELs for some of the immunologic, enzyme inducing, and reproductive effects of 2,3,7,8-TCDD in rhesus monkeys are below 1 ng/kg/day (see Table 8).

### *4.6 Pharmacokinetic Considerations in the Risk Assessment for 2,3,7,8-TCDD*

For the vast majority of chemicals, the magnitude of the toxic response is a function of the concentration of the biologically active form of the chemical at the target tissue (Klaassen, 1986). Ideally, effective doses of toxic compounds would be expressed as the concentration at the active site (delivered dose). In practice, however, doses are usually expressed as the amount of compound to which the whole organism is exposed (administered dose). Thus far in the discussion of dose-response relationships in this chapter, dose has been expressed as the amount (weight) of compound administered per kilogram of body weight. This represents a traditional approach to dose description in toxicology. So long as the concentration of the active form of the compound at the target tissue is proportional to the amount administered per unit of body weight, this is a useful convention for expressing dose and for interspecies extrapolation. This is generally the case for compounds that distribute throughout the body and do not require metabolic activation to exert their biological effect(s). For compounds that require active transport to reach the target tissue and/or require metabolic activation to exert their toxic effect, the concentration at the target tissue may not be directly proportional to the administered dose expressed as amount per unit of body weight. In such cases, factors such as the kinetic properties of the enzyme(s) responsible for the activation of the compound may be critical in determining the relationship between the amount administered and the concentration of the active form of the compound at the target tissue. Interspecies extrapolation to estimate equivalent doses for such compounds should be based on consideration of the relative activity of the metabolic enzymes in the two species. In the absence of specific information regarding relative metabolic activities in different species, it has been suggested that metabolic rate is proportional to total body surface area and therefore equivalent doses in different species should be expressed as the amount of compound administered per unit surface area (U.S. Environmental Protection Agency, 1984). Again this assumption is only valid so long as the compound is distributed and activated by the same mechanism in the two species being compared.

Although the mechanism by which 2,3,7,8-TCDD and closely related compounds cause their toxic effects has not been definitively established, the available evidence strongly suggests that many of the toxic effects including cancer promotion, teratogenesis, and immunotoxicity require, as an initial step, binding of the unmetabolized compound to the Ah receptor. The available evidence on relative rates of metabolism in different species indicates that most mammalian species metabolize 2,3,7,8-TCDD slowly and excrete the metabolites rapidly in feces and urine (Neal et al., 1984). Therefore, the relative metabolic activity of different species towards PCDDs and PCDFs is not expected to determine the relative susceptibility of those species to the toxic effects of these compounds. Interspecies differences in metabolism and elimination are not sufficient to account for all of the observed differences in acute toxicity.

The available information on the mechanism of toxicity and the metabolic fate of 2,3,7,8-TCDD in mammals suggests, but does not confirm, that the magnitude of the toxic responses caused by this and related compounds is a function of, but not totally dependent upon, the concentration of the unmetabolized compound in epithelial cells in the target organs. Studies have also shown that 2,3,7,8-TCDD is readily absorbed from the gastrointestinal tract and through the skin and is distributed into tissues high in lipid content (Neal et al., 1984; Birnbaum, 1986; Geyer et al., 1986). The retained 2,3,7,8-TCDD is eliminated very slowly and follows apparent first order kinetics, i.e., clearance varies linearly with dose. These observations suggest that the concentration of 2,3,7,8-TCDD at target tissues is a function of the whole body burden and that this body burden increases linearly as a function of administered dose. Observations of total body burdens and elimination rates have confirmed this for 2,3,7,8-TCDD in rhesus monkeys (Bowman et al., 1987b) and for 2,3,4,7,8-pentaCDF in rats (Brewster and Birnbaum, 1987).

Because of the slow rate of elimination of 2,3,7,8-TCDD and related compounds from the body, continuous exposure results in bioaccumulation and bioconcentration. At a constant rate of exposure, the body burden of PCDD/PCDF will initially increase and eventually reach a steady state concentration that will remain unchanged as long as the exposure is continued. Geyer and coworkers (1986) reviewed the available data in which concentrations of 2,3,7,8-TCDD in the diet and steady state concentrations of 2,3,7,8-TCDD in fat were determined, and they calculated bioconcentration factors (the ratio of the concentrations in the fat and the diet) from these data. Data derived from studies in rats, beef cattle, and rhesus monkeys were relatively consistent and yielded bioconcentration factors in adipose tissue ranging from 3.7 to 40. These same authors calculated a bioconcentration factor for 2,3,7,8-TCDD in humans of 153 based on a half-time for elimination of 5 years (Poiger and Schlatter, 1986).

There is a large amount of variation both among and within species in the half-life of 2,3,7,8-TCDD and closely related compounds. The half-time for elimination of 2,3,7,8-TCDD in rhesus monkeys fed 2,3,7,8-TCDD in their diet for 4 years ranged from 187–578 days with a mean of 438 (Bowman et al., 1987b). Birnbaum (1986) reported half-times of elimination of 2,3,7,8-TCDD from the adipose tissue of mice in a range of 8–13 days. For 2,3,4,7,8-pentaCDF, Brewster and Birnbaum (1987) reported a half-life of 69 days in the adipose tissue of rats. Only very limited data are available that might be used to estimate half-times for elimination of 2,3,7,8-TCDD in humans. Poiger and Schlatter (1986) measured adipose tissue levels of 2,3,7,8-TCDD in a human subject 13 and 69 days after administration of a single oral dose. They calculated a whole body half-life of 2120 days (5.8 years). Gorski et al. (1984) measured PCDD/PCDF in the adipose tissue of an adolescent female who may have been exposed to technical pentachlorophenol contaminated with highly chlorinated PCDDs. Based on measurements taken 2.5 years apart, they estimated biological half-lives of 3.5 years for 1,2,3,6,7,8-hexaCDD, 3.2 years for 1,2,3,4,6,7,8-heptaCDD, and 5.7 years for octaCDD. Recently, Pirkle and coworkers (1987) analyzed the results of measurements of 2,3,7,8-TCDD in adipose tissues at 2 separate time points in 34 individuals who were exposed to Agent Orange while assigned to Operation Ranch Hand in Vietnam. The median half-time for elimination of 2,3,7,8-TCDD from adipose tissue in these individuals was 7.1 years. The estimates of half-life from these studies are questionable because of the short time period between the measurements.

The above results suggest that 2,3,7,8-TCDD and related compounds may be more slowly eliminated from the adipose tissues of humans than they are from similar tissues in experimental animals, but this conclusion is subject to the qualification that the available data are limited. Also, calculations of elimination half-times are based on the assumption that there is no ongoing exposure to these compounds. Independent studies of adipose tissue levels of 2,3,7,8-TCDD in the general population have shown a strong correlation of tissue level with age (Graham et al., 1986; Patterson et al., 1986a; Ryan, 1986; Stanley, 1986). This implies that humans slowly accumulate 2,3,7,8-TCDD from background environmental sources over their lifetimes. Recent measurements and calculations of PCDD/PCDF in components of the human diet have led to estimates of individual daily intakes in the range of 0.015–2.0 ng of 2,3,7,8-TCDD (Graham et al., 1985; Mathar et al., 1987; Travis and Hattemer-Frey, 1987). As a result of the on-going low-level exposure in the general population and the relatively long half-time for elimination of 2,3,7,8-TCDD from the body, it is possible that steady state concentrations are never achieved. The result of this would be to increase the apparent half-time for elimination of this compound from the body as determined by serial determinations of body burdens in exposed individuals.

The study by Stanley (1986) has important implications for assessing human health risks from exposure to low levels of PCDDs/PCDFs in the environment. Forty-six composite samples of adipose tissue from the EPA National Human Adipose Tissue Survey were analyzed for a number of PCDDs and PCDFs. The exposure histories of individuals from whom the samples were taken were unknown, but overall, the samples were thought to be representative of the U.S. population at large. Measurable levels of 2,3,7,8-TCDD were found in 35 of the 46 samples and averaged 6.2 pg/g. Average adipose tissue concentrations of other PCDDs/PCDFs were 43.5 pg/g for 1,2,3,7,8-pentaCDD; 86.9 pg/g for total hexaCDD; 15.6 pg/g for 2,3,7,8-TCDF; 36.1 pg/g for 2,3,4,7,8-pentaCDF; and, 23.5 pg/g for total hexaCDF. These results confirm that humans are continually exposed to PCDDs/PCDFs in the environment and the health risks of additional exposures must be evaluated in the context of adding to an already existing body burden rather than as a *de novo* exposure.

### *4.7 Estimation of an ADI Based on Data From the Yusho and Yu-Cheng Incidents*

It has been stated above that there are no quantitative data relating exposure of humans to 2,3,7,8-TCDD to adverse health effects. There are data available, however, that do permit approximations of doses of penta- and hexachlorinated dibenzofurans that caused chloracne and related signs of intoxication in individuals who ingested cooking oils contaminated with these compounds. These data are presented in reports of studies of victims of poisonings resulting from the consumption of polychlorinated biphenyl (PCB) contaminated rice oil in Yusho, Japan and Yu-Cheng, Taiwan (see Epidemiology Working Group, this volume; and Japan-United States, 1985). Initially, the signs and symptoms of toxicity were attributed to PCBs, the major constituent of the heat-exchanger fluid. However, subsequent analysis of samples of both the heat-exchanger fluid and the contaminated rice oil showed relatively high concentrations of other chlorinated aromatic hydrocarbons which are much more biologically active in experimental test systems than are the PCB congeners that were present. Among these were a number of polychlorinated dibenzofuran isomers. On the basis of biological activity in experimental systems, their relative concentration in the contaminated rice oil, and their relative retention in the tissues of exposed victims, scientists have attributed the toxic signs and symptoms of poisoning among Yusho and Yu-Cheng victims primarily to the three PCDF congeners, 1,2,3,7,8-pentaCDF, 2,3,4,7,8-pentaCDF, and 1,2,3,4,7,8-hexaCDF (Kunita et al., 1985; Masuda et al., 1985).

Recently, Ryan and coworkers (1987) analyzed the available epidemiologic evidence regarding the consumption of contaminated rice oil by the Yusho and Yu-Cheng victims in order to estimate the total exposure to the toxic PCDF congeners among those displaying signs and symptoms of toxicity. They concluded that the smallest consumption of contaminated rice oil that was associated with the diagnosis of chloracne among Yusho victims was 2.55 ml/kg body weight corresponding to 2.32 g/kg. This amount of rice oil contained 11.6 μg of total pentaCDFs (5 ppm). This, in turn, corresponds to 1.72 μg of 2,3,4,7,8-pentaCDF equivalents based on the relative activities of the various congeners in bioassays. Assuming total gastrointestinal absorption and retention in the body, this cumulative intake would result in a body burden of 1.7 μg/kg.

For the Yu-Cheng population, data are available that relate concentrations of total pentaCDFs in the blood to the appearance of signs of intoxication (Kashimoto et al., 1983, as cited in Ryan et al., 1987). Twelve individuals with dermatologic signs had a mean blood concentration of 0.118 ng pentaCDF/g of blood. Ryan and coworkers (1987) calculated that this corresponded to 0.106 ng/g of 2,3,4,7,8-pentaCDF equivalents. Assuming that the concentration of pentaCDF in the blood is equivalent to the concentration of pentaCDF in the body on a per unit of lipid basis, these authors calculated a lipid concentration in the body of 35 μg 2,3,4,7,8-pentaCDF equivalents/kg lipid. For individuals weighing 40 kg on average and having a body fat content of 15%, the total body burden of 2,3,4,7,8-pentaCDF equivalents is 5.3 μg/kg. These two estimates of body burdens, corresponding to the development of dermatologic signs of intoxication, are quite similar.

Ryan and coworkers (1987) have used the data from the Yusho and Yu-Cheng populations to estimate a no-effect level for chloracne in exposed humans. They cited data on blood levels among victims of the Yu-Cheng incident provided by Kashimoto et al. (1983) indicating that the ratio of PCDFs in the blood of individuals with grade 3 disease (clear-cut chloracne) and those with grade 0 disease (some skin changes, but not chloracne) was approximately 3 to 1. Since the level associated with grade 0 disease might be considered to be a low-effect-level, Ryan et al. (1987) assumed that the body burden that was not associated with any disease at all would be lower by another factor of 3. Thus, they divided the 1.7 μg/kg average body burden among Yusho victims with chloracne by a factor of 9 to achieve a no-effect body burden for Yusho victims and by another factor of 10 to account for variable sensitivity within the entire human population to arrive at a no-effect level body burden for humans of 19 ng 2,3,4,7,8-pentaCDF equivalents/kg body weight. For comparison purposes, Ryan et al. (1987) used data from Stanley (1986) showing that the average background concentrations of 2,3,4,7,8-pentaCDF and total hexaCDFs in the adipose tissues of U.S. citizens are 35 pg/g and 23 pg/g, respectively. Converting these concentrations to 2,3,4,7,8-pentaCDF equivalents and using them to calculate a body burden, Ryan et al. (1987) calculated a background body burden of 8.0 ng 2,3,4,7,8-pentaCDF equivalents/kg in the U.S. population. This background body burden is just less than half of the no-effect body burden which was estimated for the Yusho data.

Using a half-time of elimination for 2,3,4,7,8-pentaCDF of 2 years derived from several studies of Yusho and Yu-Cheng vicitms, Ryan and coworkers (1987) calculated that the daily dose of 2,3,4,7,8-pentaCDF equivalents that would result in a steady state body burden of 19 ng/kg would be 18 pg/kg/day. At this rate, however, it would require over 6.5 years of exposure to achieve 90% of the steady state level and over 13 years to reach 99%. It is interesting to note that this dose rate is quite close to the ADIs for 2,3,7,8-TCDD in Table 12 which were calculated using the NOEL from the Kociba et al. (1978) study and safety factors. Studies in experimental animals indicate that 2,3,4,7,8-pentaCDF is only slightly less potent than 2,3,7,8-TCDD in causing acute effects (Safe, 1987).

The use of data from the Yusho and Yu-Cheng incidents for human health risk assessment must be qualified by several considerations. First, it was assumed that the health effects seen in the Yusho and Yu-Cheng populations were due only to the several pentaCDF and hexaCDF congeners present in the rice oil and that the presence of numerous other chlorinated compounds did not modify the quantitative toxicity of these congeners. In light of recent evidence that certain PCB isomers act as antagonists for the toxic effects of PCDDs/PCDFs that act through the Ah receptor, this assumption may be invalid (Bannister et al., 1987; Hakke et al., 1987). Second, it is generally assumed that 100% of the PCDFs in the contaminated rice oil were absorbed from the gastrointestinal tract and that the half-time for excretion of the more toxic and more presistent congeners is on the order of several years. These assumptions are reasonable based on limited data regarding the absorption, distribution, and excretion of these and related compounds in humans and non-human primates, but they have not been experimentally verified. Third, an assumption must be made regarding the relative susceptibility of humans for the chloracnegenic effects of PCDDs/PCDFs relative to other toxic effects. Data from studies in rhesus monkeys suggest that adverse reproductive effects may occur at doses below those that cause clinical chloracne (Schantz et al., 1979; Bowman et al., 1987a). Fourth, assumptions must be made about the relative biological activity of those PCDD and PCDF congeners other than the ones that were present in the contaminated rice oils. On the basis of relative activity in short-term in vitro and in vivo tests, 2,3,4,7,8-pentaCDF is nearly as active as 2,3,7,8-TCDD, but there is some uncertainty associated with extrapolating this relationship to other species, effects, and exposure regimens. Finally, the Yusho

**Table 13.** Estimated peak body burdens in experimental animals exposed to low effect levels of 2,3,7,8-tetrachlorodibenzo-*p*-dioxin[1]

| Effect | Species | Estimated Half-life | Dose and Duration | Estimated Body Burden |
|---|---|---|---|---|
| Acute lethality | Guinea pig | — | 2.0 μg/kg single dose (oral) | 2000 ng/kg |
| Cancer | Sprague-Dawley rat | 31 days[2] | 10 ng/kg/day for 2 years (in diet) | 435 (7300)[3] ng/kg |
| Cancer promotion | HRS/J hairless mouse | 11 days[4] | 7.5 ng/week for 20 weeks (dermal) | 21 (150) ng/mouse, 1100 (7900) ng/kg[5] |
| | Charles River rat | 31 days[6] | 280 ng/kg/week for 32 weeks (s.c. injection) | 1900 (9000) ng/kg |
| Chloracne | Rabbit | 31 days[6] | 4 ng/day, 5 days/week for 4 weeks (dermal) | 61 (80) ng/animal, 25 (32) ng/kg[7] |
| Chronic lethality | Rhesus monkey | 365 days[8] | 12 ng/kg/day for 9 months (in diet) | 2530 (3240) ng/kg |
| Chronic toxicity | Rhesus monkey | 365 days[8] | 1.2 ng/kg/day for 20 months (in diet) | 437 (744) ng/kg |
| Hepatic effects: | | | | |
| Altered porphyrin metabolism | Sprague-Dawley rat | 31 days[2] | 10 ng/kg/day for 2 years (in diet) | 435 (7300) ng/kg |
| Increased lipid peroxidation | Sprague-Dawley rat | 31 days[2] | 1 μg/kg single dose (oral) | 1000 ng/kg |
| Lipid accumulation | Sprague-Dawley rat | 31 days[2] | 100 ng/kg/day for 13 weeks (in diet) | 3800 (9100) ng/kg |
| Ultrastructural changes | Sprague-Dawley rat | 31 days[2] | 1 ng/kg/day for 2 years (in diet) | 43.5 (730) ng/kg |
| Immunotoxicity: | | | | |
| Alterations in T-cell subsets | Rhesus monkey | 365 days[8] | 0.12 ng/kg/day for 1 year (in diet) | 32 (43.8) ng/kg |
| Suppression of CTL | C57BL/6 mouse | 11 days[9] | 1 ng/kg/day, 1 day/week for 4 weeks (i.p. injection) | 2.3 (4) ng/kg |
| Suppression of serum complement | B6C3F1 mouse | 11 days[4] | 10 ng/kg/day for 14 days (oral) | 96 (140) ng/kg |
| Induction of AHH | Sprague-Dawley rat | 31 days[2] | 2.0 ng/kg single dose (oral) | 2.0 ng/kg |
| | Sprague-Dawley rat | 31 days[2] | 10 ng/kg/day, 2 days/week for 22 weeks (oral) | 115 (440) ng/kg |

**Table 13.** *(Continued)*

| Effect | Species | Estimated Half-life | Dose and Duration | Estimated Body Burden |
|---|---|---|---|---|
| Reproductive effects: | | | | |
| Impaired conception, spontaneous abortions | Sprague-Dawley rat | 31 days[2] | 10 ng/kg/day for 1 year (in diet) | 434 (3650) ng/kg |
| | Rhesus monkey | 365 days[8] | 0.12 ng/kg/day for 6 months to 4 years (in diet) | 19 (22) ng/kg (6 months), 59 (175) ng/kg (4 years) |
| Perinatal toxicity: | | | | |
| Decreased thymus weight | Wistar rat | 31 days[6] | 250 ng/kg to dams on day 16 of gestation (oral) | 250 ng/kg[10] |
| Induced hepatic AHH and ECOD | Wistar rat | 31 days[6] | 100 ng/kg to dams on day 16 of gestation (oral) | 100 ng/kg[10] |
| Teratogenesis: | | | | |
| Cleft palate | CF-1 mouse | 11 days[4] | 1 μg/kg/day to dams on days 6–15 of gestation (oral) | 7656 (10,000) ng/kg[10] |
| | Rhesus monkey | 365 days[8] | 1 μg/kg/day to dams on days 20–40 of gestation (oral) | 19,500 (20,000) ng/kg[10] |
| Kidney anomalies | Sprague-Dawley rat | 31 days[2] | 0.5 μg/kg/day to dams on days 6–15 of gestation (s.c. injection) | 4400 (5000) ng/kg[10] |
| | C57BL/6 mouse | 11 days[9] | 1 μg/kg/day to dams on days 10–13 of gestation (oral) | 3700 (4000) ng/kg[10] |
| | Sprague-Dawley rat | 31 days[2] | 10 ng/kg/day for 1 year prior to mating (in diet) | 434 (3650) ng/kg[10] |

[1] The body burden was calculated by the Panel for the time immediately following the last dose, using the methods described by Rose et al. (1976). 100% absorption is assumed for oral dosing; [2] Value from Rose et al. (1976); [3] Values in parentheses indicate the body burden if there was no metabolism and/or elimination of 2,3,7,8-TCDD; [4] Values assume same rate of elimination as the C57BL/6 mouse; [5] Assumes the body weight of the mice to be 20 g; [6] Assumes the same rate of elimination as the Sprague-Dawley rat; [7] Assumes the body weight of the rabbits to be 2.5 kg; [8] Value from McNulty et al. (1982); [9] Value from Gasiewicz et al. (1983); [10] Assumes that the fetal body burden is the same as the maternal, and that pregnancy does not alter the elimination rate

and Yu-Cheng incidents were the results of relatively short-term high-level exposure and the effects seen must be considered as primarily acute toxic effects. The populations are too small and the ascertainment of exposure too imprecise for these populations to serve as indicators of possible subtle and delayed toxic effects of long-term low-level exposure to pentaCDFs and related compounds. For these reasons the available data on Yusho and Yu-Cheng victims are not useful, in and of themselves, for estimating ADIs for exposure to pentaCDFs and related compounds. However, they are useful in setting a probable upper bound on the NOEL for the acute toxic effects of 2,3,7,8-TCDD after short-term exposure.

It is possible to estimate body burdens of 2,3,7,8-TCDD that are associated with biological effects in experimental animals, if one assumes first order kinetics of elimination. Table 13 shows estimated body burdens per kilogram of total body weight that are associated with the low effect levels in Table 8. These estimated body burdens range from 2.0 ng/kg for enzyme induction and suppression of cytotoxic T lymphocytes to approximately 20,000 ng/kg for induction of birth defects in rhesus monkeys. However, estimated body burdens associated with a number of effects in several species are below 100 ng/kg. These include enzyme induction, chloracne, immune alteration, impaired reproduction, and histopathologic changes in the liver. For comparison purposes the body burdens of 2,3,4,7,8-pentaCDF equivalents in Yusho and Yu-Cheng victims with chloracne were estimated to be 1,700 to 5,300 ng/kg (Ryan et al., 1987). Assuming that 2,3,4,7,8-pentaCDF has one-tenth the biological activity of 2,3,7,8-TCDD (U.S. Environmental Protection Agency, 1987), these body burdens correspond to body burdens of 170–530 ng 2,3,7,8-TCDD equivalents/kg body weight. Also using the data of Stanley (1986) for background levels of PCDDs and PCDFs in human adipose tissues and converting these to 2,3,7,8-TCDD equivalents with the EPA equivalency factors (U.S. Environmental Protection Agency, 1987), one can calculate a background average adipose tissue concentration of 39 ng 2,3,7;8-TCDD equivalents/kg adipose tissue in the U.S. population. If the average body fat composition of an adult human is 15%, then the average body burden of 2,3,7,8-TCDD equivalents is 5.8 ng/kg of body weight.

## 5 Summary and Conclusions

Studies in animals suggest that the most likely adverse human health effects that might result from chronic exposure to PCDDs/PCDFs would be cancer, adverse reproductive effects, and alterations in immune function. These effects have been demonstrated reproducibly in a wide range of animal species, but none have been found convincingly to occur in humans due to exposure to 2,3,7,8-TCDD or other PCDDs/PCDFs. Epidemiological studies, however, cannot rule out the possibility that adverse effects could occur in humans at exposures comparable to those that cause these effects in experimental animals. Limited experimental evidence from studies of human cells and tissues exposed to PCDDs/PCDFs in vitro suggests that the responses are qualitatively similar to those produced in cells and tissues from experimental animals.

While there are epidemiologic studies of humans exposed to PCDDs/PCDFs, they are not useful for quantitative risk assessment because they do not provide consistent evidence of cancer, reproductive, or immune system effects. Therefore, it is necessary to rely on quantitative dose-response data from animal studies to estimate these risks. Several scientifically supportable methods are available for quantitatively estimating human health risks from exposure to environmental concentrations of PCDDs/PCDFs using quantitative dose-response data from studies in experimental animals. However, no single method can be selected as the most appropriate on the basis of scientific criteria alone. One traditional approach for estimating human health risks from exposure to PCDDs/PCDFs is to select cancer as the endpoint of concern and to use a non-threshold mathematical model to extrapolate dose-response curves to environmentally-relevant doses from the range of responses observed in studies in experimental animals. This approach is based on the assumptions that cancer is the result of irreversible genetic alterations that are self-replicating and that doses that cause these alterations are below those that cause other toxic responses.

Another approach that has been proposed is to select a NOEL for the most sensitive adverse response elicited by PCDDs/PCDFs in experimental animals and to apply a safety factor in order to determine an acceptable daily intake for humans. Proponents of this approach cite evidence that PCDDs/PCDFs are non-genotoxic carcinogens and hypothesize that the mechanism by which these compounds "promote" a carcinogenic response involves reversible steps for which there is a threshold concentration below which they will not occur. While the weight of scientific evidence indicates that 2,3,7,8-TCDD is not a genotoxic carcinogen and does not form adducts with rat liver DNA, the precise mechanism by which it does lead to an increased incidence of cancer in experimental animals has not been elucidated. Thus, the Panel cannot conclude that the NOEL-safety factor approach is more appropriate than the non-threshold mathematical extrapolation approach for estimating human cancer risks from exposure to PCDDs/PCDFs.

It is important to realize that neither of the two general approaches to risk assessment provide a single estimate of a "safe" exposure level. Rather, both approaches provide a relatively wide range of overlapping estimates, with the estimates being highly dependent upon the toxic endpoint, the experimental dose-response data, the mathematical model or safety factor, and the scaling factors selected for calculating equivalent human doses. None of these selections can be justified on a purely scientific basis.

Recognizing that there is no single, scientifically supportable method for quantifying precise human health risks from low-level exposure to

PCDDs/PCDFs, the Panel is aware that decisions need to be made regarding the human health hazards resulting from exposures to these compounds. The weight of scientific evidence does support certain generalizations regarding the assessment and, at least, permits a narrowing of the wide range of estimates of a "safe" daily exposure. Pharmacokinetic and mechanistic data strongly suggest that the interaction of 2,3,7,8-TCDD with a stereospecific receptor protein in cells is a necessary step in eliciting many of the biological responses observed in mammals, including humans. Included among the effects that appear to be mediated by this receptor in experimental animals are: 1) promotion of cancer, 2) induction of birth defects (cleft palate), and 3) alteration of immune function. Altered gene expression may play a role in 2,3,7,8-TCDD's ability to elicit these responses. Evidence for the commonality of mechanism by which 2,3,7,8-TCDD causes many of its biological effects is provided by the observation that the lowest observed effect levels for many of these effects are similar. In contrast, many genotoxic carcinogens (e.g. polycyclic aromatic hydrocarbons and nitrosamines) induce cancer at doses considerably below those that elicit other toxic effects.

Pharmacokinetic data on 2,3,7,8-TCDD and closely related compounds in mammals indicate that these compounds are readily absorbed from the gastrointestinal tract, partially sequestered in tissue lipids, slowly metabolized, and slowly eliminated via first order kinetics. There appears to be large inter- and intraspecies variability in the half-times of elimination. Under conditions of sustained exposure, these compounds accumulate in adipose tissues and other lipid depots. Data on the levels of PCDDs/PCDFs in human adipose tissue indicate that there is a finite background exposure to these compounds in the general population and that body burdens of these compounds increase with age.

Very limited data from experiments in animals indicate that the magnitude of the biological response elicited by these compounds as a result of sustained low-level exposure correlates more closely with total body burden of the compound than with the daily dose. Therefore, assessments of human health risks from exposure to these compounds need to consider the relationship between biological responses and total body burden. This is not to say that equivalent responses will occur at equivalent body burdens in different species. Other factors, such as the ligand binding characteristics of the receptor and factors that modulate the activity of the ligand-receptor complex in regulating gene expression, may play a role in inter- and intraspecies variability in responsiveness to these compounds.

Based on a weight-of-evidence analysis of the scientific data, the Panel is of the view that linearized mathematical modeling procedures may not be appropriate for estimating PCDDs/PCDFs risks at low exposure levels. On the other hand, a traditional safety factor approach may not be warranted because the available information on mechanisms is not adequate to show that the effects of PCDDs/PCDFs are reversible. Furthermore, if a threshold exists, it is more properly expressed in terms of body burden rather than as a daily dose. A growing body of evidence indicates that populations of modern industrialized nations including the United States are continually exposed to PCDDs/PCDFs with diet being the most important source. Thus, there exist background body burdens of these compounds in humans, and the average of these is close to calculated body burdens associated with adverse effect levels in experimental animals. While the Panel is not aware of any evidence that average background body burdens in the U.S. population are at or near levels associated with adverse health effects, it is possible, because of inter-individual variability both in dietary exposure and susceptibility, that some individuals are at risk of adverse health consequences from *any* additional exposure to these compounds. In such situations, a threshold model is not useful for estimating incremental health risks from additional sources of exposure.

In order to more accurately estimate the human health risks from exposure to PCDDs/PCDFs in the environment, it is necessary to have more definitive information on the quantitative relationship between body burden and the magnitude of the biological response(s) of concern, on the relationship between daily intake and body burden, and on the relative susceptibility of humans and experimental animals to the toxic effects of these compounds. The Panel is of the view that pharmacokinetically-based risk assessment procedures that estimate incremental risks associated with body burden may be appropriate for recommending VSDs for PCDDs/PCDFs. Until such information is available it is not possible to recommend any single model for risk assessment.

Since the endpoints for cancer, reproduction, and immune effects occur at comparable dose levels in the animal studies, the Panel concludes that exposure limits protective for the carcinogenic effect would also be protective for the reproductive or immune endpoints. This conclusion is based on the assumption that the body burden is the result of a relatively constant low level exposure. In situations where variable exposures are possible, short-term exposure limits may need to be established to protect against effects other than cancer.

## 6 References


Abernethy DJ, Boreiko CJ (1987) Promotion of C3H/10 T1/2 morphological transformation by polychlorinated dibenzo-*p*-dioxin isomers. Carcinogenesis 8: 1485–1490

Abernethy DJ, Greenlee WF, Huband JC, Boreiko CJ (1985) 2,3,7,8-Tetrachlorodibenzo-*p*-dioxin (TCDD) promotes the transformation of C3H/10T1/2 cells. Carcinogenesis 6: 651–653

Allen JR, Barsotti DA, Van Miller JP, Abrahamson LJ, Lalich JJ (1977) Morphological changes in monkeys consuming a diet containing low levels of 2,3,7,8-tetrachloro-dibenzo-*p*-dioxin. Food Cosmet. Toxicol. 15: 401–410

Altman NH, Goodman DG (1979) Neoplastic diseases. In: Baker HJ, Lindsey JR and Weisbroth SH (eds) The laboratory rat. vol 1, Academic, New York, pp 334–375

Andrews JS Jr, Garrett WA Jr, Patterson DG Jr, Needham LL, Anderson JE, Roberts DW, Bagby JR, Palletta FX Sr, Palletta FX Jr (1987) 2,3,7,8-Tetrachlorodibenzo-*p*-dioxin levels in adipose tissue of exposed and control persons in Missouri. Presented at Dioxin '87: Seventh International Symposium on Chlorinated Dioxins and Related Compounds, October 4–9, Las Vegas. p 111. [Abstract]. Available from: [University of Nevada, Las Vegas]

Bannister R, Davis D, Zacharewski T, Tizard I, Safe S (1987) Aroclor 1254 as a 2,3,7,8-tetrachlorodibenzo-*p*-dioxin antagonist: effects on enzyme induction and immunotoxicity. Toxicology 46: 29–42

Berry DL, DiGiovanni J, Juchau MR, Bracken WM, Gleason GL, Slaga TJ (1978) Lack of tumor-promoting ability of certain environmental chemicals in a two-stage mouse skin tumorigenesis assay. Res. Commun. Chem. Pathol. Pharmacol. 20: 101–108

Birnbaum LS (1986) Distribution and excretion of 2,3,7,8-tetrachlorodibenzo-*p*-dioxin in congenic strains of mice which differ at the *Ah* locus. Drug Metab. Dispos. 14: 34–40

Boreiko CJ, Abernethy DJ, Sanchez JH, Dorman BH (1986) Effect of mouse skin tumor promoters upon [3*H*]uridine exchange and focus formation in cultures of C3H/10T1/2 mouse fibroblasts. Carcinogenesis 7: 1095–1099

Boreiko CJ, Abernethy DJ, Stedman DB (1987) Alterations of intercellular communication associated with the transformation of C3H/10T1/2 cells. Carcinogenesis 8: 321–325

Bowman RE, Schantz SL, Gross ML, Ferguson SX (1987a) Behavioral effects in monkeys exposed to 2,3,7,8-TCDD transmitted maternally during gestation and for four months of nursing. Presented at Dioxin '87: Seventh International Symposium on Chlorinated Dioxins and Related Compounds, October 4–9, Las Vegas. [Abstract]. Available from: [University of Nevada, Las Vegas]

Bowman RE, Schantz SL, Weerasinghe NCA, Gross ML, Barsotti DA (1987b) Chronic dietary intake of 2,3,7,8-tetrachlorodibenzo-*p*-dioxin (TCDD) at 5 and 25 parts per trillion in the monkey: TCDD kinetics and dose-effect estimate of reproductive toxicity. Presented at Dioxin '87: Seventh International Symposium on Chlorinated Dioxins and Related Compounds, October 4–9, Las Vegas. [Abstract]. Available from: [University of Nevada, Las Vegas]

Brewster DW, Birnbaum LS (1987) Disposition and excretion of 2,3,4,7,8-pentachlorodibenzofuran in the rat. Toxicol. Appl. Pharmacol. 90: 243–252

Brewster DW, Matsumura F, Akera T (1987) Effects of 2,3,7,8-tetrachlorodibenzo-*p*-dioxin on guinea pig heart muscle. Toxicol. Appl. Pharmacol. 89: 408–417

Butterworth BE, Slaga TJ (eds) (1987) Nongenotoxic mechanisms in carcinogenesis. Banbury Report 25. Cold Spring Harbor, NY: Cold Spring Harbor Laboratory, 397 p

Clark DA, Gauldie J, Szewczuk MR, Sweeney G (1981) Enhanced suppressor cell activity as a mechanism of immunosuppression by 2,3,7,8-tetrachlorodibenzo-*p*-dioxin. Proc. Soc. Exp. Biol. Med. 168: 290–299

Clark DA, Sweeney G, Safe S, Hancock E, Kilburn DG, Gauldie J (1983) Cellular and genetic basis for suppression of cytotoxic T cell generation by haloaromatic hydrocarbons. Immunopharmacology 6: 143–153

Cohen GM, Bracken WM, Iyer RP, Berry DL, Selkirk JK, Slaga TJ (1979) Anticarcinogenic effects of 2,3,7,8-tetrachlorodibenzo-*p*-dioxin on benzo(a)pyrene and 7,12-dimethylbenz(a)anthracene tumor initiation and its relationship to DNA binding. Cancer Res. 39: 4027–4033

Cordle F (1981) The use of epidemiology in the regulation of dioxins in the food supply. Regul. Toxicol. Pharmacol. 1: 379–387

Courtney KD (1976) Mouse teratology studies with chlorodibenzo-*p*-dioxins. Bull. Environ. Contam. Toxicol. 16: 674–681

Courtney KD, Moore JA (1971) Teratology studies with 2,4,5-trichlorophenoxyacetic acid and 2,3,7,8-tetrachlorodibenzo-*p*-dioxin. Toxicol. Appl. Pharmacol. 20: 396–403

Dean JH, Luster MI, Boorman GA, Chae K, Lauer LD, Luebke RW, Lawson LD, Wilson RE (1981) Assessment of immunotoxicity induced by the environmental chemicals 2,3,7,8-tetrachlorodibenzo-*p*-dioxin, diethylstilbestrol, and benzo(a)pyrene. In: Hadden J, Chedid L, Mullen P and Spreafico F (eds) Advances in immunopharmacology. Pergamon, New York, pp 37–50

Dean JH, Murray MJ, Ward EC (1986) Toxic responses of the immune system. In: Klaassen CD, Amdur MO and Doull J (eds) Casarett and Doull's toxicology: the basic science of poisons. 3rd edn, Macmillan, New York, pp 245–285

DiGiovanni J, Viaje A, Berry DL, Slaga TJ, Juchau MR (1977) Tumor initiating ability of 2,3,7,8-tetrachlorodibenzo-*p*-dioxin (TCDD) and Arochlor 1254 in a two-stage system of mouse skin carcinogenesis. Bull. Environ. Contam. Toxicol. 18: 552–557

DiGiovanni J, Juchau MR, Berry DL, Slaga TJ (1979) 2,3,7,8-Tetrachlorodibenzo-*p*-dioxin: potent anti-carcinogenic activity in CD-1 mice. Biochem. Biophys. Res. Commun. 86: 577–584

Faith RE, Luster MI (1979) Investigations on the effects of 2,3,7,8-tetrachlorodibenzo-*p*-dioxin (TCDD) on parameters of various immune functions. Ann. N.Y. Acad. Sci. 320: 564–571

Faith RE, Moore JA (1977) Impairment of thymus-dependent immune functions by exposure of the developing immune system to 2,3,7,8-tetrachlorodibenzo-*p*-dioxin (TCDD). J. Toxicol. Environ. Health 3: 451–465

Gasiewicz TA, Geiger LH, Rucci G, Neal RA (1983) Distribution, excretion, and metabolism of 2,3,7,8-tetrachlorodibenzo-*p*-dioxin in C57BL/6J, DBA/2J, and B6D2F1/J mice. Drug Metab. Disp. 11: 397–403

Geyer HJ, Scheunert I, Korte F (1986) Bioconcentration potential of organic environmental chemicals in humans. Regul. Toxicol. Pharmacol. 6: 313–347

Giavini E, Prati M, Vismara C (1982a) Rabbit teratology study with 2,3,7,8-tetrachlorodibenzo-*p*-dioxin. Environ. Res. 27: 74–78

Giavini E, Prati M, Vismara C (1982b) Effects of 2,3,7,8-tetrachlorodibenzo-*p*-dioxin administered to pregnant rats during the preimplantation period. Environ. Res. 29: 185–189

Goldstein JA, Graham MJ, Sloop T, Maronpot R, Goodrow T, Lucier GW (1987) Effects of 2,3,7,8-tetrachlorodibenzo-*p*-dioxin (TCDD) on enzyme-altered foci, hepatocellular tumors, and estradiol metabolism in a two-stage carcinogenesis model. Presented at Dioxin '87: Seventh International Symposium on Chlorinated Dioxins and Related Compounds, October 4–9, Las Vegas. [Abstract]. Available from: [University of Nevada, Las Vegas]

Gorski T, Konopka L, Brodzki M (1984) [Persistence of some polychlorinated dibenzo-*p*-dioxins and polychlorinated dibenzofurans of pentachlorophenol in human adipose tissue]. Rocz. Panstw. Zakl. Hig. 35: 297–301

Graham M, Hileman F, Kirk D, Wendling J, Wilson J (1985) Background human exposure to 2,3,7,8-TCDD. Chemosphere 14: 925–928

Graham M, Hileman FD, Orht RG, Wendling JM, Wilson JD (1986) Chlorocarbons in adipose tissue from a Missouri population. Chemosphere 15: 1595–1600

Greenlee WF, Dold KM, Osborne R (1985a) Actions of 2,3,7,8-tetrachlorodibenzo-*p*-dioxin (TCDD) on human epidermal keratinocytes in culture. In Vitro Cell Dev. Biol. 21: 509–512

Greenlee WF, Dold KM, Irons RD, Osborne R (1985b) Evidence for direct action of 2,3,7,8-tetrachlorodibenzo-*p*-dioxin (TCDD) on thymic epithelium. Toxicol. Appl. Pharmacol. 79: 112–120

Haake JM, Safe S, Mayura K, Phillips TD (1987) Aroclor 1254 as an antagonist of the teratogenicity of 2,3,7,8-tetrachlorodibenzo-*p*-dioxin. Toxicol. Lett. 38: 299–306

Hassoun EM, Dencker L (1982) TCDD embryotoxicity in the mouse may be enhanced by beta-napthoflavone, another ligand of the Ah-receptor. Toxicol. Lett. 12: 191–198

Hong R, Taylor K, Abonour R (1987) Immune abnormalities associated with chronic TCDD exposure in rhesus. Presented at Dioxin '87: Seventh International Symposium on Chlorinated Dioxins and Related Compounds, October 4–9, Las Vegas. [Abstract]. Available from: [University of Nevada, Las Vegas]

Japan-United States (1985) Joint Seminar on Toxicity of Chlorinated Biphenyls, Dibenzofurans, Dibenzodioxins, and Related Compounds. April 25–28, 1983, Fukuoka, Japan. Environ. Health Perspect. 59: 1–181

Jones EL, Krizek HA (1962) A technic for testing acnegenic potency in rabbits applied to the potent acnegen 2,3,7,8-tetrachlorodibenzo-*p*-dioxin. J. Invest. Dermatol. 39: 511–517

Kahn PC, Gochfeld M, Nygren M, Hansson M, Rappe C, Velez H, Ghent-Guenther T, Wilson WP (1988) Dioxins and dibenzofurans in blood and adipose tissue of Agent Orange-exposed Vietnam veterans and matched controls. J. Am. Med. Assoc. 259: 1661–1667

Kashimoto T, Miyata H, Fukashima H, Kunita N (1983) [Study on PCBs, PCQs, and PCDFs in the blood of Taiwanese patients with PCB poisoning and the causal cooking of rice-bran oil]. Fukuoka Acta Med. 74: 255–268

Kelling CK, Menahan LA, Peterson RE (1986) Influence of isoproterenol on tension development and rate in atria isolated from rats treated with 2,3,7,8-tetrachlorodibenzo-*p*-dioxin (TCDD). Toxicologist 6: 312 [Abstract]

Khera KS, Ruddick JA (1973) Polychlorodibenzo-*p*-dioxin: perinatal effects and the dominant lethal test in Wistar rats. In: Blair EH (ed) Chlorodioxins origin and fate. Advances in Chemistry Series 120. Washington, DC: American Chemical Society

Kimbrough RD, Falk H, Stehr P, Fries G (1984) Health implications of 2,3,7,8-tetrachlorodibenzodioxin (TCDD) contamination of residential soil. J. Toxicol. Environ. Health 14: 47–93

Kitchin KT, Woods JS (1978) 2,3,7,8-Tetrachlorodibenzo-*p*-dioxin induction of aryl hydrocarbon hydroxylase in female rat liver: evidence for de novo synthesis of cytochrome P-448. Mol. Pharmacol. 14: 890–899

Klaassen CD (1986) Distribution, excretion, and absorption of toxicants. In: Klaassen CD, Amdur MO, Doull J (eds) Casarett and Doull's toxicology: the basic science of poisons. 3rd edn, Macmillan, New York, pp 33–63

Knutson JC, Poland A (1984a) 2,3,7,8-Tetrachlorodibenzo-*p*-dioxin: examination of the biochemical effects involved in the proliferation and differentiation of XB cells. J. Cell. Physiol. 121: 143–151

Knutson JC, Poland A (1984b) XB cells: an in vitro model for the differentiation and proliferation response to 2,3,7,8-TCDD. In: Poland A, Kimbrough RD (eds) Banbury Report 18. Biological mechanisms of dioxin action. Cold Spring Harbor, NY: Cold Spring Harbor Laboratory, pp 381–389

Kociba RJ, Keeler PA, Park CN, Gehring PJ (1976) 2,3,7,8-Tetrachlorodibenzo-*p*-dioxin (TCDD): results of a 13-week oral toxicity study in rats. Toxicol. Appl. Pharmacol. 35: 553–574

Kociba RJ, Keyes DG, Beyer JE, Carreon RM, Wade CE, Dittenber DA, Kalnins RP, Franson LE, Park CN, Bernard SD, Hummel RA, Humiston CG (1978) Results of a two-year chronic toxicity and oncogenicity study of 2,3,7,8-tetrachlorodibenzo-*p*-dioxin in rats. Toxicol. Appl. Pharmacol. 46: 279–303

Kouri RE, Rude TH, Joglekar R, Dansette PM, Jerina DM, Atlas SA, Owens IS, Nebert DW (1978) 2,3,7,8-Tetrachlorodibenzo-*p*-dioxin as cocarcinogen causing 3-methylcholanthrene-initiated subcutaneous tumors in mice genetically "nonresponsive" at *Ah* locus. Cancer Res. 38: 2777–2783

Krewski D, Brown C, Murdoch D (1984) Determining "safe" levels of exposure: safety factors or mathematical models? Fundam. Appl. Toxicol. 4: S383–S394

Kunita N, Hori S, Obana H, Otake T, Nishimura H, Kashimoto T, Ikegami N (1985) Biological effect of PCBs, PCQs and PCDFs present in the oil causing Yusho and Yu-Cheng. Environ. Health Perspect. 59: 79–84

Lesca P (1983) Modulating effects of 2,3,7,8-tetrachlorodibenzo-*p*-dioxin on skin carcinogenesis initiated by 7,12-dimethylbenz(a)anthracene in CF-1 Swiss mice. In: Rydstroem J, Montelius J and Bengtsson M (eds) Extrahepatic drug metabolism and chemical carcinogenesis. Proceedings of an International Meeting on Extrahepatic Drug Metabolism and Chemical Carcinogenesis, May 17–20, Stockholm, Sweden. Elsevier, Amsterdam, pp 589–590

Luster MI, Dean JH, Boorman GA (1982) Altered immune functions in rodents treated with 2,3,7,8-tetrachlorodibenzo-*p*-dioxin, phorbol-12-myristate-13-acetate, and benzo(a)pyrene. In: Hunt V, Smith MK, Worth D (eds) Banbury Report 11. Environmental factors in human growth and development. Cold Spring Harbor, NY: Cold Spring Harbor Laboratory, pp 199–215

Luster MI, Hong LH, Boorman GA, Clark G, Hayes HT, Greenlee WF, Dold K, Tucker AN (1985) Acute myelotoxic responses in mice exposed to 2,3,7,8-tetrachlorodibenzo-*p*-dioxin (TCDD). Toxicol. Appl. Pharmacol. 81: 156–165

Madsen C, Larsen JC (1987) Relative toxicity of chlorinated dibenzo-*p*-dioxins and dibenzofurans measured by thymus weight and liver enzyme induction in perinatally dosed rats: 2,3,7,8-TCDD, 2,3,4,7,8-PeCDF and 1,2,3,7,8-PeCDF. Presented at Dioxin '87: Seventh International Symposium on Chlorinated Dioxins and Related Compounds, October 4–9, Las Vegas. [Abstract]. Available from: [University of Nevada, Las Vegas]

Masuda Y, Kuroki H, Haraguchi K, Nagayama J (1985) PCB and PCDF congeners in the blood and tissues of Yusho and Yu-Cheng patients. Environ. Health Perspect. 59: 53–58

Mathar W, Beck H, Eckhart K, Ruhl CS, Wittkowski R (1987) Body burden with PCDDs and PCDFs from food intake in Germany. Presented at Dioxin '87: Seventh International Symposium on Chlorinated Dioxins and Related Compounds, October 4–9, Las Vegas. [Abstract]. Available from: [University of Nevada, Las Vegas]

McConnell EE (1985) The clinicopathologic changes in various species of animals caused by dibenzo-*p*-dioxins. In: Kamrin MA, Rodgers PW (eds) Dioxins in the environment. Hemisphere, Washington, DC, pp 225–230

McConnell EE, Moore JA, Dalgard DW (1978) Toxicity of 2,3,7,8-tetrachlorodibenzo-*p*-dioxin in rhesus monkeys (*Macacca mulatta*) following a single oral dose. Toxicol. Appl. Pharmacol. 43: 175–187

McNulty WP (1977) Toxicity of 2,3,7,8-tetrachlorodibenzo-*p*-dioxin for rhesus monkeys: brief report. Bull. Environ. Contam. Toxicol. 18: 108–109

McNulty WP (1984) Fetotoxicity of 2,3,7,8-tetrachlorodibenzo-*p*-dioxin (TCDD) for rhesus macaques (*Macacca mulatta*). Am. J. Primatol. 6: 41–47

McNulty WP, Pomerantz I, Farrell T (1981) Chronic toxicity of 2,3,7,8-tetrachlorodibenzofuran for rhesus macaques. Food Cosmet. Toxicol. 19: 57–65

McNulty WP, Nielsen-Smith KA, Lay JO Jr, Lippstreu DL, Kangas NL, Lyon PA, Gross ML (1982) Persistence of TCDD in monkey adipose tissue. Food Cosmet. Toxicol. 20: 985–987

Moore JA, Gupta BN, Zinkl JG, Vos JG (1973) Postnatal effects of maternal exposure to 2,3,7,8-tetrachlorodibenzo-*p*-dioxin (TCDD). Environ. Health Perspect. 5: 81–85

Murray FJ, Smith FA, Nitschke KD, Humiston CG, Kociba RJ, Schwetz BA (1979) Three generation reproduction study of rats given 2,3,7,8-tetrachlorodibenzo-*p*-dioxin (TCDD) in the diet. Toxicol. Appl. Pharmacol. 50: 241–252

Nagarkatti PS, Sweeney GD, Gauldie J, Clark DA (1984) Sensitivity to suppression of cytotoxic T cell generation by 2,3,7,8-tetrachlorodibenzo-*p*-dioxin (TCDD) is dependent on the *Ah* genotype of the murine host. Toxicol. Appl. Pharmacol. 72: 169–176

Nagayama J, Kiyohara C, Masuda Y, Kuratsune M (1985) Genetically mediated induction of aryl hydrocarbon hydroxylase in human lymphoblastoid cells by polychlorinated dibenzofuran isomers and 2,3,7,8-tetrachlorodibenzo-*p*-dioxin. Arch. Toxicol. 56: 230–235

National Cancer Institute (1979) Bioassay of 2,7-dichlorodibenzo-*p*-dioxin (DCDD) for possible carcinogenicity. NCI Carcinogenesis Technical Report Series No. 123. Washington, DC

National Research Council (1980) Drinking water and health. vol 3. Washington, DC: National Academy Press, pp 25–65

National Toxicology Program (1980a) Carcinogenesis bioassay of 1,2,3,6,7,8- and 1,2,3,7,8,9-hexachlorodibenzo-*p*-dioxin for possible carcinogenicity (gavage study). DHHS Publication No. (NIH) 80-1754. Available from: National Technical Information Service, Springfield, VA

National Toxicology Program (1980b) Carcinogenesis bioassay of 1,2,3,6,7,8- and 1,2,3,7,8,9-hexachlorodibenzo-*p*-dioxin for possible carcinogenicity (dermal study). DHHS Publication No. (NIH) 80-1758. Available from: National Technical Information Service, Springfield, VA

National Toxicology Program (1982a) Carcinogenesis bioassay of 2,3,7,8-tetrachlorodibenzo-*p*-dioxin (CAS No. 1746-01-6) in Osborne-Mendel rats and B6C3F1 mice (gavage study). DHHS Publication No. (NIH) 82-1765. Available from: National Technical Information Service, Springfield, VA

National Toxicology Program (1982b) Carcinogenesis bioassay of 2,3,7,8-tetrachlorodibenzo-*p*-dioxin (CAS No. 1746-01-6) in Swiss-Webster mice (dermal study). DHHS Publication No. (NIH) 82-1757. Available from: National Technical Information Service, Springfield, VA

Nau H, Bass R (1981) Transfer of 2,3,7,8-tetrachlorodibenzo-*p*-dioxin (TCDD) to the mouse embryo and fetus. Toxicology 20: 299–308

Neal RA (1985) Mechanisms of the biological effects of PCBs, polychlorinated dibenzo-*p*-dioxins and polychlorinated dibenzofurans in experimental animals. Environ. Health Perspect. 60: 41–46

Neal R, Gasiewicz TA, Geiger L, Olson J, Sawahata T (1984) Metabolism of 2,3,7,8-tetrachlorodibenzo-*p*-dioxin in mammalian species. In: Poland A, Kimbrough RD (eds) Banbury Report 18. Biological mechanisms of dioxin action. Cold Spring Harbor, NY: Cold Spring Harbor Laboratory. pp 49–60

Neubert D, Dillman I (1972) Embryotoxic effects in mice treated with 2,4,5-trichlorophenoxyacetic acid and 2,3,7,8-tetrachlorodibenzo-*p*-dioxin. Arch. Pharmacol. 272: 243–264

Nisbet ICT, Paxton MB (1982) Statistical aspects of three-generation studies of the reproductive toxicity of TCDD and 2,4,5-T. Am. Stat. 36: 290–298

Ontario. Ministry of the Environment (1985) Polychlorinated dibenzo-*p*-dioxins (PCDDs) and polychlorinated dibenzofurans (PCDFs). Scientific Criteria Document for Standard Development No. 4-84. Available from: [The Ministry]

Patterson DG, Holler JS, Smith SJ, Liddle JA, Sampson EJ, Neeham LL (1986a) Human adipose tissue data for 2,3,7,8-tetrachlorodibenzo-*p*-dioxin in certain U.S. samples. Chemosphere 15: 2055–2060

Patterson DG Jr, Hoffman RE, Needham LL, Roberts DW, Bagby JR, Pirkle JL, Falk H, Sampson EJ, Houk VN (1986b) 2,3,7,8-Tetrachlorodibenzo-*p*-dioxin levels in adipose tissue of exposed and control persons in Missouri: an interim report. J. Am. Med. Assoc. 256: 2683–2686

Pirkle JL, Wolff WH, Patterson DG, Needham LL, Michael JE, Miner JC, Peterson MR (1987) Estimates of the half-life of 2,3,7,8-tetrachlorodibenzo-*p*-dioxin in Ranch Hand veterans. Presented at Dioxin '87: Seventh International Symposium on Chlorinated Dioxins and Related Compounds, October 4–9, Las Vegas. [Abstract]. Available from: [University of Nevada, Las Vegas]

Pitot HC, Goldsworthy T, Campbell HA, Poland A (1980) Quantitative evaluation of the promotion by 2,3,7,8-tetrachlorodibenzo-*p*-dioxin of hepatocarcinogenesis from diethylnitrosamine. Cancer Res. 40: 3616–3620

Pitot HC, Goldsworthy T, Moran S (1981) The natural history of carcinogenesis: implications of experimental carcinogenesis in the genesis of human cancer. J. Supramol. Struct. Cell Biochem. 17: 133–146

Poiger H, Schlatter C (1986) Pharmacokinetics of 2,3,7,8-TCDD in man. Chemosphere 15: 1489–1494

Poland A (1984) Reflections on the mechanism of action of halogenated aromatic hydrocarbons. In: Poland A and Kimbrough RD (eds) Biological mechanisms of dioxin action. Banbury Report 18. Cold Spring Harbor, NY: Cold Spring Harbor Laboratory, pp 109–117

Poland A, Knutson JC (1982) 2,3,7,8-Tetrachlorodibenzo-*p*-dioxin and related halogenated aromatic hydrocarbons: examination of the mechanism of toxicity. Annu. Rev. Pharmacol. Toxicol. 22: 517–554

Poland A, Palen D, Glover E (1982) Tumor promotion by TCDD in skin of HRS/J hairless mice. Nature (London) 300: 271–273

Poland A, Knutson J, Glover E, Kende A (1983) Tumor promotion in the skin of hairless mice by halogenated aromatic hydrocarbons. In: Weinstein IB, Vogel HJ (eds) Genes and proteins in oncogenesis. Academic, New York, pp 143–161

Portier CJ, Hoel DG, Van Ryzin J (1984) Statistical analysis of the carcinogenesis bioassay data relating to the risks from exposure to 2,3,7,8-tetrachlorodibenzo-*p*-dioxin. In: Lowrance WW (ed) Public health risks of the dioxins. New York: Rockefeller University, pp 99–118

Puhvel SM, Sakamoto M (1987) Responses of murine epidermal keratinocyte cultures to 2,3,7,8-tetrachlorodibenzo-*p*-dioxin (TCDD): comparison of haired and hairless genotypes. Toxicol. Appl. Pharmacol. 89: 29–36

Puhvel SM, Reisner RM, Ertl DC (1985) Effect of TCDD on murine epidermal keratinocytes in vitro. Chemosphere 14: 971–973

Rice RH, Cline PR (1984) Response of malignant epidermal keratinocytes to 2,3,7,8-TCDD. In: Poland A, Kimbrough RD (eds) Banbury Report 18. Biological mechanisms of dioxin action. Cold Spring Harbor, NY: Cold Spring Harbor Laboratory. pp 373–380

Roberts EA, Golas CL, Okey AB (1986) Ah receptor mediating induction of aryl hydrocarbon hydroxylase: detection in human lung by binding of 2,3,7,8-[*3H*]tetrachlorodibenzo-*p*-dioxin. Cancer Res. 46: 3739–3743

Rose JQ, Ramsey JC, Wentzler TH, Hummel RA, Gehring PJ (1976) The fate of 2,3,7,8-tetrachlorodibenzo-*p*-dioxin following single and repeated oral doses to the rat. Toxicol. Appl. Pharmacol. 36: 209–226

Ryan JJ (1986) Variations of dioxins and furans in human tissues. Chemosphere 15: 1585–1593

Ryan JJ, Mukerjee D, Brown JF Jr, Gasiewicz TA (1987) Estimation of total body burden of PCDFs associated with chloracne in humans based on the Yusho and Yu-Cheng poisonings. Presented at Dioxin '87: Seventh International Symposium on Chlorinated Dioxins and Related Compounds, October 4–9, Las Vegas. [Abstract]. Available from: [University of Nevada, Las Vegas]

Safe SH (1986) Comparative toxicology and mechanism of action of polychlorinated dibenzo-*p*-dioxins and dibenzofurans. Annu. Rev. Pharmacol. Toxicol. 26: 371–399

Safe S (1987) Determination of the 2,3,7,8-TCDD toxic equivalent factors: support for the use of the in vitro AHH induction assay. Chemosphere 16: 791–802

Saffiotti U (1977) Scientific bases of environmental carcinogenesis and cancer prevention: developing an interdisciplinary science and facing its ethical implications. J. Toxicol. Environ. Health 2: 1435–1447

Schantz SL, Barsotti DA, Allen JR (1979) Toxicological effects produced in nonhuman primates chronically exposed to fifty parts per trillion 2,3,7,8-tetrachlorodibenzo-*p*-dioxin (TCDD). Toxicol. Appl. Pharmacol. 48: A180 [Abstract]

Schiller CM, King MW, Walden R (1986) Alterations in lipid parameters associated with changes in 2,3,7,8-tetrachlorodibenzo-*p*-dioxin (TCDD)-induced mortality in rats. In: Rappe C, Choudhary GJ, Keith LH (eds) Chlorinated dioxins and dibenzofurans in perspective. Lewis, Chelsea, MI, pp 285–302

Schwetz BA, Norris JM, Sparschu GL, Rowe UK, Gehring PJ, Emerson JL, Gerbig CG (1973) Toxicology of chlorinated dibenzo-*p*-dioxins. Environ. Health Perspect. 5: 87–99

Sharma RP, Gehring PJ (1979) Effects of 2,3,7,8-tetrachlorodibenzo-*p*-dioxin (TCDD) on splenic lymphocyte transformation in mice after single and repeated exposures. Ann. N.Y. Acad. Sci. 320: 487–497

Shu HP, Paustenbach DJ, Murray FJ (1987) A critical evaluation of the use of mutagenesis, carcinogenesis, and tumor promotion data in a cancer risk assessment of 2,3,7,8-tetrachlorodibenzo-*p*-dioxin. Regul. Toxicol. Pharmacol. 7: 57–88

Sielken RL Jr (1987) Quantitative cancer risk assessments for 2,3,7,8-tetrachlorodibenzo-*p*-dioxin (TCDD). Food Chem. Toxicol. 25: 257–267

Sleight SD, Jensen RK, Rezabek MS (1987) Enhancement of hepatocarcinogenesis in rats by simultaneous administration of 2,4,5,2',4',5'-hexachlorobiphenyl (HCB) and 2,3,7,8-tetrachlorodibenzo-*p*-dioxin (TCDD). Toxicologist 7: 103

Sloop TC, Lucier GW (1987) Dose-dependent elevation of Ah receptor binding by TCDD in rat liver. Toxicol. Appl. Pharmacol. 88: 329–337

Smith FA, Schwetz BA, Nitschke KD (1976) Teratogenicity of 2,3,7,8-tetrachlorodibenzo-*p*-dioxin in CF-1 mice. Toxicol. Appl. Pharmacol. 38: 517–523

Society of Toxicology (1982) Animal data in hazard evaluation: paths and pitfalls. Fundam. Appl. Toxicol. 2: 101–107

Sparschu GL, Dunn FL, Rowe VK (1971) Study of the teratogenicity of 2,3,7,8-tetrachlorodibenzo-*p*-dioxin in the rat. Food Cosmet. Toxicol. 9: 405–412

Stanley JS (1986) Broad scan analysis of human adipose tissue. vol 4. Polychlorinated dibenzo-*p*-dioxins (PCDDs) and polychlorinated dibenzofurans (PCDFs). Final report prepared for the Office of Toxic Substances of the U.S. Environmental Protection Agency under Contract No. 68-02-4252. EPA 560/5-86-038. 56 p. Available from: National Technical Information Service, Springfield, VA

Stohs SJ, Al-Bayati ZF, Hassan MQ, Murray WJ, Mohammadpour HA (1986) Glutathione peroxidase and reactive oxygen species in TCDD-induced lipid peroxidation. Adv. Exp. Med. Biol. 197: 357–365

Sweeney G, Barford D, Rowley B, Goddard G (1984) Mechanisms underlying the hepatotoxicity of 2,3,7,8-tetrachlorodibenzo-*p*-dioxin. In: Poland A, Kimbrough RD (eds) Banbury Report 18. Biological mechansims of dioxin action. Cold Spring Harbor, NY: Cold Spring Harbor Laboratory. pp 225–239

Thigpen JE, Faith RE, McConnell EE, Moore JA (1975) Increased susceptibility to bacterial infection as a sequela of exposure to 2,3,7,8-tetrachlorodibenzo-*p*-dioxin. Infect. Immunol. 12: 1319–1324

Thomas PT, Hinsdill RD (1979) The effect of perinatal exposure to tetrachlorodibenzo-*p*-dioxin on the immune response of young mice. Drug Chem. Toxicol. 2: 77–98

Thunberg T (1984) Effect of TCDD on vitamin A and its relation to TCDD toxicity. In: Poland A, Kimbrough RD (eds) Banbury Report 18. Biological mechanisms of dioxin action. Cold Spring Harbor, NY: Cold Spring Harbor Laboratory. pp 333–344

Toth K, Sugar J, Somfai-Relle S, Bence J (1978) Carcinogenic bioassay of the herbicide 2,4,5-trichlorophenoxy-ethanol (TCPE) with different 2,3,7,8-tetrachlorodibenzo-*p*-dioxin (dioxin) content in Swiss mice. Prog. Biochem. Pharmacol. 14: 82–93

Toth K, Somfai-Relle S, Sugar J, Bence J (1979) Carcinogenicity testing of herbicide 2,4,5-trichlorophenoxyethanol containing dioxin and of pure dioxin in Swiss mice. Nature (London) 278: 548–549

Travis CC, Hattemer-Frey HA (1987) Human exposure to 2,3,7,8-TCDD. Chemosphere 16: 2331–2342

U.S. Environmental Protection Agency (1984) Proposed guidelines for carcinogen risk assessment. Fed. Regist. 49: 46294–46301

U.S. Environmental Protection Agency (1985) Health assessment document for polychlorinated dibenzo-*p*-dioxins. Office of Health and Environmental Assessment. 600/8-84/014F. Washington, DC. Available from: National Technical Information Service, Springfield, VA

U.S. Environmental Protection Agency (1987) Interim procedures for estimating risks associated with exposures to mixtures of chlorinated dibenzo-*p*-dioxins and dibenzofurans (CDDs and CDFs). EPA/625/3-87/012. Washington, DC. Available from: National Technical Information Service, Springfield, VA

U.S. Office of Science and Technology Policy (1985) Chemical carcinogens: a review of the science and its associated principles. Fed. Regist. 50: 10372–10442

van der Heijden CA, Knaap AGAC, Kramers PGN, Van Logten MJ (1982) Evaluation of the carcinogenicity and mutagenicity of 2,3,7,8-tetrachlorodibenzo-1,4-dioxin (TCDD): classification and no-effect level. Report DOC/LCM 300/292. State Institute of National Health, Bilthoven, The Netherlands.

Van Miller JP, Lalich JJ, Allen JR (1977) Increased incidence of neoplasms in rats exposed to low levels of 2,3,7,8-tetrachlorodibenzo-*p*-dioxin. Chemosphere 9: 537–544

Vecchi A, Sironi M, Canegrati MA, Recchia M, Garattini S (1983) Immunosuppressive effects of 2,3,7,8-tetrachlorodibenzo-*p*-dioxin in strains of mice with different susceptibility to induction of aryl hydrocarbon hydroxylase. Toxicol. Appl. Pharmacol. 68: 434–441

Vecchi A, Graziani A, Sironi M, Dal Fiume D, Sfreddo-Gallotta E, Saletti MC, Cantoni L (1985) Simultaneous administration of TCDD and TCDF at different ratios induces different effects. Chemosphere 14: 957–961

Vos JG, Moore JA (1974) Suppression of cellular immunity in rats and mice by maternal treatment with 2,3,7,8-tetrachlorodibenzo-*p*-dioxin. Int. Arch. Allergy Appl. Immunol. 47: 777–794

Vos JG, Moore JA, Zinkl JG (1973) Effect of 2,3,7,8-tetrachlorodibenzo-*p*-dioxin on the immune system of laboratory animals. Environ. Health Perspect. 5: 149–162

Vos JG, Kreeftenberg JG, Engel HW, Minderhoud A, Van Noorle Jansen LM (1978) Studies on 2,3,7,8-tetrachlorodibenzo-*p*-dioxin induced immune suppression and decreased resistance to infection: endotoxin hyper-sensitivity, serum zinc concentrations and effect of thymosin treatment. Toxicology 9: 75–86

Weber G, Luzi P, Resi L, Tanganelli P, Lovati MR, Poli A (1983) Natural history of TCDD-induced liver lesions in rats as observed by transmission electron microscopy during a 32-week period after a single intraperitoneal injection. J. Toxicol. Environ. Health 12: 533–540

White KL Jr, Lysy HH, McCay JA, Anderson AC (1986) Modulation of serum complement levels following exposure to polychlorinated dibenzo-*p*-dioxins. Toxicol. Appl. Pharmacol. 84: 209–219

Wong TK, Sloop T, Lucier GW (1986) Nondetectable concentrations of human placental Ah receptors are associated with potent induction of microsomal benzo[a]pyrene hydroxylase in individuals exposed to polychlorinated biphenyls, quaterphenyls, and dibenzofurans. Toxicol. Appl. Pharmacol. 85: 60–68

Zingeser MR (1979) Anomalous development of the soft palate in rhesus macaques (*Macacca mulatta*) prenatally exposed to 2,3,7,8-tetrachlorodibenzo-*p*-dioxin. Teratology 19: 54A–55A [Abstract]

# Subject Index


Acceptable daily intake estimation
  PCBs, contaminated rice oil 124–126, 129
    methodology 115
  2,3,7,8-tetrachlorodibenzo-*p*-dioxin 121
ADIs, see Acceptable daily intake
Agent Orange
  birth defects 36–40
  sarcomas, soft tissue, risks 43, 44–46
  spraying, aerial 37
Agriculture
  herbicides
    sarcomas, soft tissue, risks 42–43
Analysis
  polychlorinated dibenzofurans in the atmosphere 11–12
  polychlorinated dibenzo-*p*-dioxins in the atmosphere 11–12
Anergy
  2,3,7,8-tetrachlorodibenzo-*p*-dioxin 34–36
Aroclor 1010
  contamination by PCDFs 7
Aroclor 1242
  contamination by PCDFs 5, 7
Aroclor 1254
  contamination by PCDFs 7
Aroclor 1260
  contamination by PCDFs 7
Atmosphere
  polychlorinated dibenzofurans analysis 11–12
  polychlorinated dibenzo-*p*-dioxin analysis 11–12
Automobiles
  emissions of PCDDs and PCDFs 10

Binding
  2,3,7,8-tetrachlorodibenzo-*p*-dioxin receptors 64–83
Birth defects
  Agent Orange 36–40
    New Zealand 39
  2,4-D 39
  2,3,7,8-tetrachlorodibenzo-*p*-dioxin
    Italy 38–39
    West Virginia 40

Cancer
  herbicides – phenoxy, incidence 50–54
  2,3,7,8-tetrachlorodibenzo-*p*-dioxin incidence, animals 120
    risk assessment 117–119
Carcinogenicity
  2,7-dichlorodibenzo-*p*-dioxin in rodents 99
  dioxins 40
  furans 40
  1,2,3,6,7,8-hexachlorodibenzo-*p*-dioxin in rodents 100
  1,2,3,7,8,9-hexachlrodibenzo-*p*-dioxin in rodents 100
  2,3,7,8-tetrachlorodibenzo-*p*-dioxin
    dose response relationship 110–114
    mixtures 101–103
    risk assessment 116–132
    rodents 97–98
Chemical industry
  lymphomas, malignant – risk 49
  trichlorophenol – sarcomas, soft tissue 46
Chloracne
  epidemiology 30–31
  2,3,7,8-tetrachlorodibenzo-*p*-dioxin 40
Chlorophenols
  sarcomas, soft tissue – risks 41–42
Cholesterol levels
  2,3,7,8-tetrachlorodibenzo-*p*-dioxin effects 32–33
Contamination
  Aroclor 1010
    by PCDFs 7
  Aroclor 1242
    by PCDFs 5, 7
  Aroclor 1254
    by PCDFs 7

Aroclor 1260
by PCDFs 7
Dowicide G
by PCDDs 4–5;
by PCDFs 4–5
fish
Great Lakes 13–15
pentachlorophenols
by PCDDs and PCDFs 4–5
rice oil
by PCBs 7, 16–17, 30
by PCBs, ADI estimation 124–125, 129
sediments
by PCDDs and PCDFs 12
2,4-D birth defects 39
sarcomas, soft tissue – risks 44
2,7-Dichlorodibenzo-*p*-dioxin
carcinogenicity in rodents 99

Dioxins
epidemiology 24–53
flue gases 8
fly ash 8
paper 8
paper recycled 8
shellfish 8
Dosage
2,3,7,8-tetrachlorodibenzo-*p*-dioxin and toxicity 96–114
Dowicide G
contamination
by PCDDs 4–5
by PCDFs 4–5

Emissions
automobiles
PCDDs and PCDFs 10
flue gases
dioxins 8
incinerators 8–9
polychlorinated dibenzofurans in the atmosphere, analysis 11–12
polychlorinated dibenzo-*p*-dioxins in the atmosphere, analysis 11–12
Environmental impact
polychlorinated dibenzofurans 11
polychlorinated dibenzo-*p*-dioxins 11
Epidemiology
cancer
risks, phenoxy herbicides 50–54
chloracne 30–31
dioxins 24–54
lymphomas, malignant 47–49
lymphomas, non-Hodgkins 48–49
myelomas, multiple 47
polychlorinated dibenzofurans 27–29
polychlorinated dibenzo-*p*-dioxins 27–29
sarcomas, soft tissue 47

Fires
polychlorinated biphenyls as sources of PCDDs and PCDFs 9–10
Fish
contamination
by PCDDs 12–15
by PCDFs 12–15
by 2,3,7,8-tetrachlorodibenzo-*p*-dioxin 12–15
Great Lakes 14–15
Flue gases
dioxins 8
Fly ash
dioxins 8
Food poisoning
rice oil contamination, by PCBs 7
Forestry herbicides
sarcomas, soft tissue, risks 42–43

Gasoline, leaded
emissions of PCDDs and PCDFs 10
Great Lakes
fish contamination 12–14
polychlorinated dibenzofurans
contamination, fish 12–14
polychlorinated dibenzo-*p*-dioxins
contamination, fish 12–14
2,3,7,8-tetrachlorodibenzo-*p*-dioxin fish contamination, 12–15

Herbicides
2,4-D
birth defects 39
Agent Orange
spraying, effects 38–40, 43
agriculture
sarcomas, soft tissue, risks 42–43
forestry
sarcomas, soft tissue, risks 42–43
phenoxy
cancer, incidence 50–54
lymphomas, malignant 47–49
lymphomas, non-Hodgkins 47–49
phenoxyacetic acid
spraying, risks 41, 43
Hexachlorobenzene contamination
by PCDFs and PCDDs 7
Hexachlorocyclohexanes contamination
by PCDDs and PCDFs 7
1,2,3,6,7,8-Hexachlorodibenzo-*p*-dioxin
carcinogenicity in rodents 100
1,2,3,7,8,9-Hexachlorodibenzo-*p*-dioxin
carcinogenicity in rodents 100

Immune system
2,3,7,8-tetrachlorodibenzo-*p*-dioxin effects 34–35
Incinerators
emissions
PCDDs 8–9
PCDFs 8–9
Isomers
polychlorinated dibenzofurans 1–2
polychlorinated dibenzo-*p*-dioxins 1–2
Italy
sarcomas, soft tissue
incidence, Seveso 46–47
2,3,7,8-tetrachlorodibenzo-*p*-dioxin
birth defects 38–39

Kanechlor 400
contamination by PCDFs 5

Lake Huron
fish contamination 13–14
Lake Michigan
fish contamination 13–14
Lake St. Clare
fish contamination 13–14
Lake Superior
fish contamination 13–14
Lipophilicity
polychlorinated dibenzofurans 1–2
polychlorinated dibenzo-*p*-dioxins 1–2
Liver function tests
2,3,7,8-tetrachlorodibenzo-*p*-dioxin changes 32, 36
Lymphomas
2,3,7,8-tetrachlorodibenzo-*p*-dioxin 40–41
Lymphomas, malignant
epidemiology 47–49
Lymphomas, non-Hodgkins
epidemiology 48–49

Mechanism of action
polychlorinated dibenzofurans toxicity 61–83
polychlorinated dibenzo-*p*-dioxins toxicity 61–83
2,3,7,8-tetrachlorodibenzo-*p*-dioxin
binding, receptors 64–72
role, Ah receptor 75–78
toxicity 122–124
Methodology
acceptable daily intakes 115
2,3,7,8-tetrachlorodibenzo-*p*-dioxin 121
toxicity risk assessment 115–132
virtually safe doses 115
Models
2,3,7,8-tetrachlorodibenzo-*p*-dioxin toxicity 80–83
polychlorinated dibenzofurans toxicity 65
polychlorinated dibenzo-*p*-dioxins toxicity 65
toxicity risk assessment 115–132
Mutagenicity
polychlorinated dibenzofurns DNA damage 64
polychlorinated dibenzo-*p*-dioxins
DNA damage 64
Myelomas, multiple
epidemiology
international 47–48

Neurotoxicity
2,3,7,8-tetrachlorodibenzo-*p*-dioxin 35–36
New Zealand
Agent Orange birth defects 39
2,4-D birth defects 39

Paper – dioxins 8
Paper, recycled – dioxins 8
PCBs, see Polychlorinated biphenyls
PCDDs, see Polychlorinated dibenzofurans
PCDFs, see Polychlorinated dibenzo-*p*-dioxins
Pentachlorophenols
contamination by PCDDs and PCDFs 4–5
Pharmacogenetics
Ah receptors 80
Pharmacokinetics
2,3,7,8-tetrachlorodibenzo-*p*-dioxin
use, risk assessment 121–124
Phenoxyacetic acid
spraying risks 41, 43
PolyCDDs, see Polychlorinated dibenzo-*p*-dioxins
PolyCDFs, see Polychlorinated dibenzofurans
Polychlorinated biphenyls
chloracne
epidemiology 30–31
contamination
by PCDFs 6
rice oil 7, 16–17, 30, 124–126, 129
fires
sources of PCDDs and PCDFs 9–10
tissue levels 16–17
Polychlorinated dibenzofurans
chloracne epidemiology 30–31
contamination
of hexachlorobenzene 7
of hexachlorocyclohexanes 7
of Kanechlor 400, 5
of PCBs 6
sediments 12
sludges 9, 12
sources, gasoline, leaded, 10

Polychlorinated dibenzofurans (*Contd.*)
emissions
atmosphere, analysis 11–12
epidemiology 24–54
mechanism of action
toxicity 61–83
models
toxicity 65
mutagenicity
DNA damage 64
pulp and paper industry
sources 7–8, 11
risk assessment 129–132
sources
chemical 4
fires, PCBS 9–10
incinerators 8–9
industrial 3
structure 1–4
structure-toxicity relationships 78–80
tissue levels 16–17
toxicity
human health risks 93–132
Polychlorinated dibenzo-*p*-dioxins
chloracne
epidemiology 30–31
contamination
of hexachlorobenzene 7
of hexachlorocyclohexanes 7
of sediments 12
of sludges 9, 12
emissions
atmosphere, analysis 11–12
epidemiology 24–54, 27–29
human health risks 93–132
mechanism of action
toxicity 61–83
models
toxicity 65
mutagenicity
DNA damage 64
risk assessment 129–132
sources
chemical 4
contamination, leaded gas 10
fires, PCBs 9–10
incinerators 8–9
industrial 3
structure 1–4
structure-toxicity relationships 78–80
tissue levels 16–17
toxicity
human health risks 93–132
Porphyria
polychlorinated dibenzofurans 31–32
polychlorinated dibenzo-*p*-dioxins 31–32
Porphyria cutania tarda
2,3,7,8-tetrachlorodibenzo-*p*-dioxin 32
Pulp and paper industry
polychlorinated dibenzofurans sources 7–8, 11
polychlorinated dibenzo-*p*-dioxins sources 7–8, 11

Receptors, Ah
binding of 2,3,7,8-tetrachlorodibenzo-*p*-dioxin 64–83
pharmacogenetics 80
Reproduction
Agent Orange effects 37–38
2,3,7,8-tetrachlorodibenzo-*p*-dioxin effects 37–40, 104
Rice oil
contamination by PCBs 7, 16–17, 30, 124–126, 129
Risk assessment
methodology 115–132
polychlorinated dibenzofurans
toxicity 129–132
polychlorinated dibenzo-*p*-dioxins
toxicity 129–132
2,3,7,8-tetrachlorodibenzo-*p*-dioxin 116–132
pharmacokinetics 121–124

Sarcomas, soft tissue
2,4-D – risks 44
Agent Orange – risks 43–46
chemical industry 46
chlorophenols – risks 41–42
epidemiology 47
Italy
incidence, Sevesa 46–47
phenoxyacetic acid – risks 41–43
2,3,7,8-tetrachlorodibenzo-*p*-dioxin 40–42
trichlorophenol 46
Sediments
contamination by PCDDs and PCDFs 12
Shellfish
dioxins 8
Sludge
contamination by PCDDs and PCDFs 9, 12
Spraying, aerial
Agent Orange – effects, reproduction 36–37
Stability
polychlorinated dibenzofurans 1–2
polychlorinated dibenzo-*p*-dioxins 1–2
Structure
polychlorinated dibenzofurans 1–3

Structure-toxicity relationships
  polychlorinated dibenzofurans 78–80
  polychlorinated dibenzo-*p*-dioxins 78–80
Sweden
  lymphomas, malignant – epidemiology 47–48

2,4,5-T, see Agent Orange
TCDD, see 2,3,7,8-Tetrachlorodibenzo-*p*-dioxin
2,3,7,8-TCDD, see 2,3,7,8-Tetrachlorodibenzo-*p*-dioxin
Teratogenicity
  2,3,7,8-tetrachlorodibenzo-*p*-dioxin in rodents 104
2,3,7,8-Tetrachlorodibenzo-*p*-dioxin
  acceptable daily intakes – methodology 121
  anergy 34–36
  birth defects
    Italy 38–39
    Vietnam 37–38
    West Virginia 40
  body burden in animals 127–129
  cancer incidence, effects, animals 120
  carcinogenicity
    dose response relationships 110–114
    mixtures 101–103
    risk assessment 116–132
    rodents 97–98
  chloracne 30–31, 40
  cholesterol levels and effects 32, 33
  contamination of PCBs 7
  dosage and toxicity 96–114
  fish contamination, Great Lakes 12–15
  immune system effects 34–35
  liver function tests – changes 32, 36
  lymphomas 40–41
  mechanism of action
    role, Ah receptor 75–78
    toxicity 122–124
    toxicity, receptor binding 64–72
  models of toxicity 80–83
  neurotoxicity 35–36
  pharmacokinetics, use, risk assessment 121–124
  porphyria cutania tarda 32
  receptors, Ah – binding 69–83
  reproduction effects 37–40
  sarcomas, soft tissue 40–42
  teratogenicity in rodents 104
  toxicity
    evaluation 95–110
    human health risks 93–132
    immune function, rodents 106–108
  triglyceride levels and effects 32–33
  ulcers 34, 36
Tissue levels
  polychlorinated dibenzofurans 16–17
  polychlorinated dibenzo-*p*-dioxin 16–17
Toxicity
  methodology of risk assessment 115–132
  models of risk assessment 115–132
  polychlorinated dibenzofurans
    humans health risks 93–132
    mechanism of action 61–83
    risk assessment 129–132
  polychlorinated dibenzo-*p*-dioxins
    human health risks 93–132
    mechanism of action 61–83
    risk assessment 129–132
  2,3,7,8-tetrachlorodibenzo-*p*-dioxin
    evaluation 95–110
    models 80–83
    immune function, rodents 106–108
Trichlorophenol
  sarcomas, soft tissue 46
2,4,5-Trichlorophenoxy acetic acid, see Agent Orange
Triglyceride levels
  2,3,7,8-tetrachlorodibenzo-*p*-dioxin, effects 32–33

Ulcers
  2,3,7,8-tetrachlorodibenzo-*p*-dioxin, effects 34, 36
United States
  lymphomas, malignant – incidence 48–49
  lymphomas, non-Hodgkins – incidence 48–49

Vietnam
  Agent Orange
    effects, reproduction 37–38
    sarcomas, soft tissue 44–46
  2,3,7,8-tetrachlorodibenzo-*p*-dioxin, birth defects 37–38
Virtually safe doses
  2,3,7,8-tetrachlorodibenzo-*p*-dioxin 117–118
  methodology 115
VSDs, see Virtually safe doses

West Virginia
  2,3,7,8-tetrachlorodibenzo-*p*-dioxin, birth defects 40

Yu-Cheng, see Rice oil
Yusho, see Rice oil

## Archives of Environmental Contamination and Toxicology



An international, interdisciplinary journal of full-length articles, **Archives of Environmental Contamination and Toxicology** covers original experimental and theoretical research on this subject. Detailed reports of significant advances and discoveries in the fields of air, water, and soil contamination and pollution are published. In addition, research results from disciplines concerned with the introduction, presence, and effects of deleterious substances in the total environment are documented in **Archives of Environmental Contamination and Toxicology.**

ISSN 0090-4341 Title No. 244

| Subscription Information 1990 | | |
|---|---|---|
| Vol. 19 | 6 issues | Price DM 640,- |
| Plus carriage charges: FRG DM 26,32<br>Other countries DM 41,10 | | |

**Springer-Verlag New York**

## Bulletin of Environmental Contamination and Toxicology



Dedicated to the rapid publication of camera-ready results in the field, **Bulletin of Environmental Contamination and Toxicology** disseminates advances and discoveries in the areas of air, soil, water, and food contamination and pollution. Descriptions of methods, procedures, and techniques are designed to allow readers to apply them to their own laboratory work. Through articles which are briefer than those found in archival journals, the **Bulletin of Environmental Contamination and Toxicology** provides a meeting ground for researchers to share in new discoveries as soon as they are made.

ISSN 0007-4861 Title No. 128

| Subscription Information 1990 | | |
|---|---|---|
| Vol. 44+45 | 6 issues each | Price DM 581,- |
| Plus carriage charges: FRG DM 46,22<br>Other countries DM 63,- | | |

**Springer-Verlag New York**

Springer